Strategies
for Structural
Dynamic Modification

MECHANICAL ENGINEERING RESEARCH STUDIES

ENGINEERING DYNAMICS SERIES

Series Editor: **Professor J. B. Roberts,** *University of Sussex, England*

1. Feedback Design of Systems with Significant Uncertainty*
   M. J. Ashworth

2. Modal Testing: Theory and Practice
   D. J. Ewins

3. Transfer Function Techniques and Fault Location*
   J. Hywel Williams

4. Parametric Random Vibration
   R. A. Ibrahim

5. Statistical Dynamics of Nonlinear and Time-Varying Systems
   M. F. Dimentberg

6. Strategies for Structural Dynamic Modification
   John A. Brandon

* Out of print

Strategies for Structural Dynamic Modification

John A. Brandon

School of Engineering,
University of Wales, College of Cardiff, UK

RESEARCH STUDIES PRESS LTD.
Taunton, Somerset, England

JOHN WILEY & SONS INC.
New York · Chichester · Toronto · Brisbane · Singapore

RESEARCH STUDIES PRESS LTD.
24 Belvedere Road, Taunton, Somerset, England TA1 1HD

Marketing and Distribution:

Australia and New Zealand:
Jacaranda-Wiley Ltd.
GPO Box 859, Brisbane, Queensland 4001, Australia

Canada:
JOHN WILEY & SONS CANADA LIMITED
22 Worcester Road, Rexdale, Ontario, Canada

Europe, Africa, Middle East and Japan:
JOHN WILEY & SONS LIMITED
Baffins Lane, Chichester, West Sussex, England

North and South America:
JOHN WILEY & SONS INC.
605 Third Avenue, New York, NY 10158, USA

South East Asia:
JOHN WILEY & SONS (SEA) PTE LTD.
37 Jalan Pemimpin 05-04
Block B Union Industrial Building, Singapore 2057

Library of Congress Cataloging-in-Publication Data
Brandon, John A., 1950–
    Strategies for structural dynamic modification / John A. Brandon.
      p.   cm.—(Mechanical engineering research studies)
  (Engineering dynamics series ; 6)
    Includes bibliographical references.
    ISBN 0-86380-098-X.—ISBN 0-471-92661-2 (Wiley)
    1. Structural dynamics. 2. Structural analysis (Engineering)
  I. Title. II. Series. III. Series: Engineering dynamics series ; 6.
  TA654.B67 1990
  624.1'71—dc20   89-70316
                                   CIP

British Library Cataloguing in Publication Data
Brandon, John Anthony
  Strategies for structural dynamic modification.
  1. Structure. Design
  I. Title II. Series
  624.171

  ISBN 0 86380 098 X

ISBN 0 86380 098 X (Research Studies Press Ltd.)
ISBN 0 471 92661 2 (John Wiley & Sons Inc.)

Printed in Great Britain by Galliard (Printers) Ltd., Great Yarmouth

Preface

This book is intended to describe and classify a number of methods used in structural dynamics. These techniques are rather controversial — and quite rightly so! Although powerful they are potentially dangerous when misused. They rely on the construction of quite complex models from extremely sparse data.

Procedures for assessing the effect of modifications to a structure with complete design data are well established. Typically discrepancies between the predicted and measured properties are used to adjust the design database using least squares procedures.

Less common, and less well analysed, are procedures which attempt to assess the effect of design changes using only experimental data. This would be necessary, for example, in a "troubleshooting" context where use of the full design database is impossible, either because of inaccessibility or on financial grounds.

The approach used in the book will be descriptive rather than prescriptive. It is intended that this work will provide an introduction into a relatively new area of research (although there are already a number of proprietary software packages which use these techniques). It is hoped that this volume will provide an objective

v

source of analysis and will stimulate debate as to the value of the methods discussed.

The scope of the work is covered very briefly in a section of a previous volume in this series: Modal Testing: Theory and Practice, by David Ewins. This work would provide an excellent introductory text to establish the context of the current volume which is more limited in scope.

The general rule used in the current text is to cite literature of other workers in the structural dynamics community rather than more general texts. For example the work of Meirovitch is used for reference to the finite element method rather than the more obvious general texts, for example those by Zienkiewicz or Bathe. This has the advantage that the terminology contained in the bibliography will, in general, be familiar but the quality of analysis may occasionally be lacking. In such cases reference will also be made to the general literature.

Acknowledgments

I would like to record my thanks to a large number of people who have made the current work possible.

The work originated during my time at the University of Manchester Institute of Science and Technology where Alan Cowley and Keyanoush Sadeghipour provided advice and encouragement. My thanks are due to Virgil Snyder for drawing my attention to some important work which I had overlooked.

I have received considerable benefit from the teaching and advice of David Ewins over the last twenty years. The most common comment (from him) has been: "Are you sure you know what you are letting yourself in for?" The answer has invariably been to the affirmative —— but not always with unity coherence!

I would like to thank Professor Brian Brinkworth, and my other colleagues in the School of Engineering at UWCC, for giving me the breathing space to complete the manuscript. In particular, I would like to thank Mike Green for taking much more of the burden of research administration than it is reasonable to expect.

Last, but not least, my gratitude is due to my family for their patience and tolerance.

Contents

ix

CHAPTER 1
Context

1.1   REANALYSIS: A GENERAL OVERVIEW

Structural Dynamic Modification is a term used in Modal Analysis. In the wider Computer Aided Engineering Community, for example in Finite Element Analysis, the corresponding methods are often described as Reanalysis. As will be seen, the subject matter of the present work fits rather uncomfortably into the more general definition.

The entire life cycle of a product involves an iterative engineering effort. The initial design concept is likely to give rise to a variety of design options each of which will be evaluated against the requirements of the product specification. Often at this stage a prototype or series of protypes will be constructed and their properties compared with the design idealisation.   Reanalysis, as usually understood, implies the incorporation, into an existing model, of new information gained either from experimental testing or some other source, which questions or improves the accuracy of the model.

The techniques used are usually only considered valid if the computational cost of the reanalysis is significantly less than that of the incorporation of the new information into the original model and re-solution using the original procedure or   if the solution is believed to be more accurate.   There   has   been a considerable effort to develop methods for efficient incorporation of new information into the iterative design cycle for original equipment manufacture. For most applications of interest to the dynamicist the in-service operation of the structure will also be examined, not only from the point of view of design improvements for future developments, but also because of commitments to the customer for warranties and servicing, and additionally the necessity for conformity to national standards and safety legislation.

Historically these requirements have been restricted to industries where high structural integrity has been of paramount importance, eg aerospace, boilers and pressure vessels, nuclear facilities and large civil engineering projects. With the increasing emphasis on product liability legislation the scope for in-service structural assessment is likely to widen significantly.

Consider first the most elementary type of reanalysis technique. In the Finite Element Method dynamic analysis is often undertaken using the subspace iteration technique (see for example Meirovitch(1980)). Computational advantage can be gained by using the subspace solution from an initial design study as the starting iterate for the problem of a slightly modified structure, see fig 1. This is based on the credible premise that the solution of the modified structure is likely to be close to that of the original model. It would be expected that the rate of convergence will be significantly more rapid in this case than when the trial space is chosen arbitrarily.

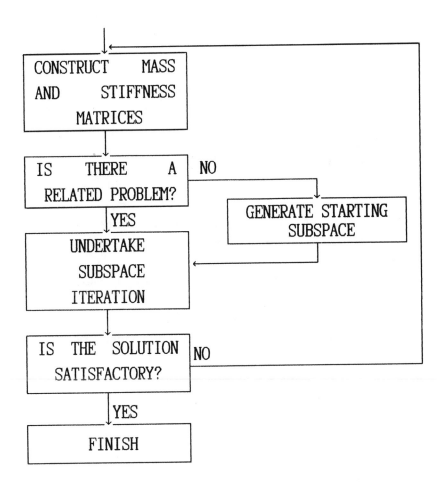

FIGURE 1: TRADITIONAL PROCEDURE FOR FINITE ELEMENT ANALYSIS

A more recent concept in Finite Element Analysis has been the development of hierarchical methods. Prior to the development of these methods the normal procedure for solving a problem of a relatively minor modification, to a system whose solution was known, involved the generation of a totally new mesh and complete re-solution of the problem. This takes place not only in response to design change, but in the Finite Element Method is implicit in the solution procedure. If a model gives unacceptable results then the entire mesh is re-defined, re-solved and reassessed, this iterative process being repeated until the solution is satisfactory. Frequently, however, the existing solution is an accurate representation throughout the majority of the structure and is unsatisfactory in only a small region, for example around a stress concentration. The reanalysis is now often achieved through a process known as adaptive mesh refinement where the change desired is restricted to a small part of the model, see fig 2.

The variation in the Finite Element mesh is isolated in its effect and the majority of the model may be carried forward unchanged into successive stages of analysis. As before, the dynamic problem, using subspace iteration, would use the final stage of one solution as the starting approximation to the next phase. Hierarchical methods are classified as h-type, if the modification entails mesh refinement, or p-type, if the modification retains the same mesh points but uses progressively more sophisticated elements. The recent literature has seen the introduction of the hybrid method, known not surprisingly as h-p refinement. The development of hierarchical methods has been described in a series of papers (by Zienkiewicz, Babuska, Meirovitch and others) in the International Journal of Numerical Methods in Engineering from 1980 onwards. Whilst these methods are outside the scope of the current work there is little doubt that they will have a major impact in Structural Dynamics. For a more general survey of reanalysis, including particularly static analysis, the reader is referred to the study by

Arora(1976).

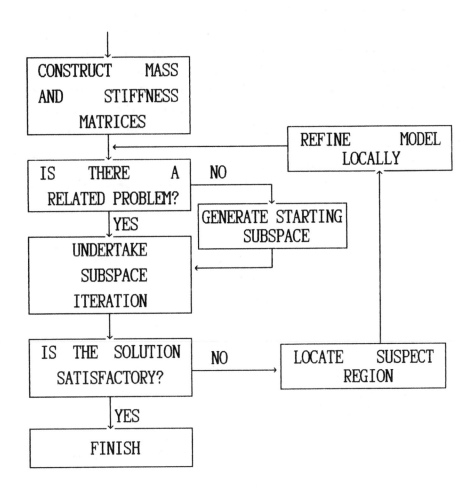

FIGURE 2: SUBSPACE ITERATION WITH HIERARCHICAL
REFINEMENT

In general the problem addressed by finite element analysts comprises only the analytical model in comparison with a design specification. The more general problem usually considered in structural dynamics has three facets. Firstly, a performance specification defines the requirements for the behaviour in service of the equipment, in terms both of required and undesirable or prohibited characteristics. Secondly, the designer's proposed solution often has to be based on a much idealised approximation of the eventual detailed structure. The third aspect is the actual behaviour measured on a model, prototype or real structure, see figure 3.

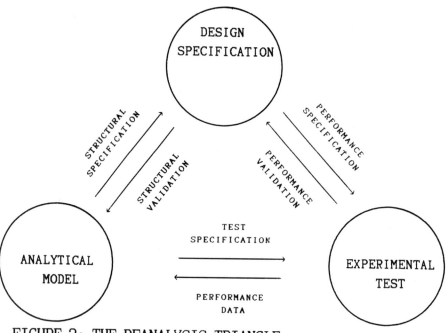

FIGURE 3: THE REANALYSIS TRIANGLE

Reanalysis in structural dynamics, therefore, addresses the problem of how to combine the available design data and the results of the experimental tests, either to demonstrate that the structure meets its specification or to indicate the required modifications to achieve it. The subject of the current work is, however, more restricted in scope but nevertheless of considerable importance. It is assumed that design data are not available, which could be due to a variety of reasons. For example, the structure of interest could have undergone a number of undocumented modifications after installation, or the original designers regarded the development of detailed models for structural dynamic assessment to be unnecessary, failing to anticipate problems in-service. This would be particularly applicable in a "troubleshooting" role. It is necessary in this context to predict the effect of structural modifications given only a (perhaps limited) set of experimental data (see fig 4).

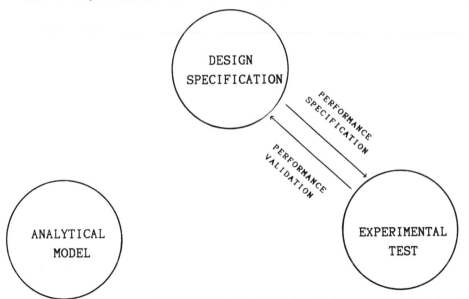

FIGURE 4: THE REANALYSIS TRIANGLE WITH EXPERIMENTAL DATA

1.2 MODEL CHARACTERISTICS IN STRUCTURAL DYNAMICS

There is a wide variety of modelling strategies used at various stages of outline specification, conceptual development, detail design and service use of structures subject to dynamic loading conditions. The analysis will be largely restricted to the assumption of viscously damped linear systems. This may appear, at first sight, to be unduly restrictive, but the methods will often be capable of being extended to other dissipative mechanisms or to certain classes of non-linear system.

Ideally a common model would be used throughout the entire life of the structure. This does not however reflect the current status quo prevailing in structural design. There are two primary reasons for this state of affairs. Firstly, individual techniques are often developed in isolation, with little consideration of the characteristics of related problems. Secondly, the techniques themselves may have inherent theoretical or practical limitations which make compatibility with other methods difficult. For example, the finite element method uses models with large numbers of degrees of freedom but is not generally regarded as being amenable to the incorporation of damping. In contrast, structural dynamics testing, typically modal analysis, is practically restricted to a relatively small number of degrees of freedom but implicitly includes assessment of damping.

It should be realised that, because of the differences between the various models, certain precedence rules apply, ie once a particular technique is used then certain other methods cannot subsequently be applied. Thus the most realistic objective is to maintain as much information as possible for as long as possible. Dynamic models, and hence reanalysis techniques, fall into two primary classes. The more general category is the modal model, describing vibration in terms of mode shapes and corresponding natural frequencies and damping ratios. Of more limited applicability, though no less important, are the response models. These characterise structural properties in terms of point to point transfer functions, for example receptance, the force-displacement transfer function. The former is an implicitly linear structural representation, whilst, as will be shown, the latter allows reanalysis approaches which include non-linear features (locally at least). The transformation between the modal and response representations is the primary cause of the irreversibility in the analysis, although it should be realised that the two approaches are essentially analytically equivalent, with the response characteristics often being identified as the initial stage of modal analysis.

1.3 HISTORICAL PERSPECTIVE

Much of the theoretical basis of the current work is long established, the foundation of modal testing theory dating to the work of Frazer, Duncan and Collar (1938). Much of the current text, the analytical basis of reanalysis, stems from work by Lancaster (1960, 1966). Perhaps the most neglected work of significance in structural dynamics is that of Kron (1963). This is in part due to the obscure notation, based on the programming techniques of analogue computers, but also because of the rather metaphysical nature of the argument. Noble (1969) attributes the appreciation of the significance of the single most important theorem in structural dynamics to Kron. This theorem will be used extensively in the current text. Kron's work has been extended by Simpson (1980) and Collar and Simpson (1987).

Further recent work, developed from Kron's methods, is

due to Sehmi (1985) and Turner et al (1985). The most comprehensive treatment of eigenvalue problems, the basis of the theory of modal analysis is that of Wilkinson (1965). A valuable companion to this volume would be the text by Bishop, Gladwell and Michaelson (1965).

The subject of reanalysis using experimental data falls into two further categories. Firstly, the methods based on exact algebra can be derived from Kron's theorem. For the modal model, important contributions are due to Weissenburger (1966, 1968), Pomazal (1969) and Pomazal and Snyder (1970). Exact methods for response analysis are characterised by the work of Mahalingam (1975a,b).

The second type are based on truncated series, typically Taylor series, and lead to methods described as sensitivity analysis. This field has (perhaps appropriately) been typified by a large number of minor advances. These methods have been analysed in considerable depth by Whitesell (1980). The subject has also been surveyed by Baldwin and Hutton (1985).

Thus the study of reanalysis methods, using only experimental data, falls into four classes, as shown in fig 5.

METHOD	EXACT	SERIES
MODAL	WEISSENBURGER AND POMAZAL	MODAL SENSITIVITY
RESPONSE	PSEUDOFORCE/ KRON'S METHOD	RESPONSE SENSITIVITY

FIGURE 5: CLASSES OF REANALYSIS METHOD

1.4 STRUCTURE OF THE WORK

Chapter 2 surveys the necessary theory for the understanding of the reanalysis problem. The remainder of the work is structured to reflect the classification shown in figure 5.

Chapter three describes exact techniques for reanalysis of the eigenvalue problem,now commonly called "Structural Dynamic Modification" (SDM). In the fourth chapter sensitivity analysis of system modal properties is described. Chapter five describes the exact reanalysis of response models using equivalent methods, developed separately, as the pseudoforce method and Kron's method. In chapter six the extension of sensitivity analysis to response methods is described. Chapter seven assesses the current potential of the methods.

CHAPTER 2
Theoretical Background

2.1 THE ORDER OF SYSTEM APPROXIMATIONS

The matrix analysis of vibration implicitly assumes a topological representation of a structure in terms of a lattice of springs, masses and dampers. Few, if any, practical structures have such a convenient topology. The work of Fredholm, however, established that a continuous structure can be represented, to any required degree of accuracy, by a discrete approximation (see for example Lanczos (1957)). Indeed the entire basis of the Finite Element Method, the Rayleigh-Ritz procedure, is founded on this premise. The continuous structure is considered in terms of a limiting case, as the order of the system matrices approaches infinity. The connections between continuum and discrete models are considered in some detail by Meirovitch (1967) (see also Meirovitch (1980), Bishop, Gladwell and Michaelson (1965) and Collar and Simpson (1987)).

Representations of the same structure, by matrices of different order, and the transformations between these models, will be of some considerable interest in this text. Of particular importance are the bracketing or inclusion theorems, which provide mutual bounds for the modal properties of different approximations for the same structure. The necessary matrix theory is covered in a number of texts, particularly Wilkinson (1965), Bishop,

14

Gladwell and Michaelson (1965), Collar and Simpson (1987) and Jennings (1977). Only material essential to the understanding of the current text will therefore be briefly reviewed here.

2.2 THE MATRIX MODEL OF A STRUCTURE

The dynamic equations of a linear structure are usually expressed in the form

$$M\,\ddot{x} + C\,\dot{x} + 1/\omega\,H\,x + K\,x = \ell(t) \qquad (2.1)$$

Where M, C, H, and K are the (nxn square, symmetric, constant) mass, viscous damping, hysteretic (or material or structural) damping and stiffness matrices respectively; x is the vector of displacements at the n points representing the discrete structure; ω is the frequency; and $\ell(t)$ the vector of external loads dependent on time t. In the form shown, the equation is a hybrid of the frequency and time domains and consequently intractable. Expressed in this form the implication is that $\ell(t)$ is harmonic, ie can be represented as a function of frequency.

In this work, in common with the majority ,of texts in structural dynamics, the analysis will be based primarily on the viscous damping model. It can be convincingly argued, however, that the hysteretic damping (or structural damping) model is more realistic representation of the behaviour of "real" engineering structures. A thorough treatment of the hysteretic damping model is provided by Bishop and Johnson (1960). The equation is often written in the alternative form

$$M\,\ddot{x} + C\,\dot{x} + (K + iH)\,x = \ell(t) \qquad (2.2)$$

a device that fails to eliminate the coupling of frequency and time, as the physical interpretation of the product iH is a time delay in terms of a phase difference of $\pi/2$, which is only meaningful for harmonic excitations. For the vast majority of practical structures, where the damping is small, the replacement of hysteretic damping by a viscous damping approximation is widely regarded as being acceptable. Other additional terms of the system matrices are used in special applications, for example, to represent gyroscopic forces (an unsymmetric frequency dependent matrix- see for example Meirovitch (1980)). These will not be analysed or used in the current text, although the methods described here are, in general, capable of extension. For the reasons discussed above the majority of the text will be restricted to the analysis of the equation

$$M \ddot{x} + C \dot{x} + K x = f(t) \tag{2.3}$$

The most common approach current in structural dynamics is Modal Analysis, based on the treatment of the homogeneous form of (2.3) as an quadratic eigenvalue problem (using the trial solution $x = \chi e^{\lambda t}$)

$$(\lambda^2 M + \lambda C + K) \chi = 0 \tag{2.4}$$

As has been mentioned in Chapter 1, for reasons which will be explained, it is quite common (and extremely advantageous) to discount the effect of damping entirely (particularly in Finite Element Analysis) and solve the problem

$$(\lambda^2 M + K) \chi = 0 \tag{2.5}$$

If necessary, damping is re-introduced using either proportional damping (qv) or a perturbation approach such as that of Lancaster (1966). Such an assumption enables analysis using only real arithmetic, whereas the inclusion of a general damping distribution not only demands the transition to complex arithmetic but also destroys advantageous numerical properties of the system matrices (in particular positive definiteness of the eigenvalue problem).

2.3 SOLUTION STRATEGIES FOR THE EIGENVALUE PROBLEM

The eigenvalue problem, equation (2.5), is a case of the generalised eigenvalue problem. The solution to such problems is most commonly approached by using a coordinate transformation to a standard eigenvalue problem $(X - \mu I) \xi = 0$. Unless the dynamic problem is poorly defined M will be positive definite, ie it is non-singular and more importantly all of its own eigenvalues are positive, and K will be at least positive semi-definite, ie, although it may be singular (if its coordinates are not sufficiently constrained), with at least one zero eigenvalue, all its non-zero eigenvalues are positive. In addition, for passive systems, C will also be at least positive semi-definite. The generalised eigenvalue problem given by equation (2.5) can be transformed, by virtue of the properties described above, to a standard eigenvalue problem simply by pre-multiplying by the inverse of M to give

$$(\lambda^2 I + M^{-1} K) \chi = 0 \qquad (2.6)$$

This transformation has the advantage that the eigenvectors are preserved, but the symmetry of the matrices is destroyed. It is generally regarded as desirable to retain symmetries of matrices wherever possible, not only to economise on storage (a symmetric matrix requires $n(n+1)/2$ rather than n) but often also to take advantage of attractive numerical properties of real symmetric matrices. A more common strategy for conversion of equation (2.5) to a standard eigenvalue problem is the use of the Choleski decomposition of the mass matrix. A real symmetric, positive-definite matrix can be factorised into the product of two triangular matrices (sometimes called Choleski square roots) which are transposes of each other

$$M = L\,L^t \qquad (2.7)$$

This factorisation is closely related to, and often used as a numerical improvement of, the common Gaussian elimination method for equation solution. The transformation of the eigenproblem using the Choleski decomposition requires a change in the coordinate system. The coordinate system denoted by the vector χ must be replaced by a system ϑ related by the coordinate transformation

$$\vartheta = L^t\,\chi \qquad (2.8)$$

or

$$\chi = L^{-t}\,\vartheta \qquad (2.9)$$

The stiffness matrix becomes

$$K = L\,X\,L^t \qquad (2.10)$$

or

$$X = L^{-1}\,K\,L^{-t} \qquad (2.11)$$

with X symmetric for symmetric K.

This allows the direct replacement of the generalised eigenvalue problem (2.5) with the standard symmetric eigenvalue problem (2.6) with the coordinate relationship between the eigenvectors defined by equations (2.8) and (2.9).

2.4 SOLVING THE QUADRATIC EIGENVALUE PROBLEM- DUNCAN'S METHOD

This method was analysed by Frazer, Duncan and Collar (1938) and as the state space method forms the basis of much of modern control theory. The method is based on replacing the n x n quadratic eigenvalue problem with a 2n x 2n linear eigenproblem. By adjoining the identity

$$M\,\dot{x}\ -\ M\,\dot{x}\ =\ 0$$

and using the coordinate transformation

$$y\ =\ \begin{pmatrix} \dot{x} \\ x \end{pmatrix}$$

equation (2.3) can be re-written

$$\begin{pmatrix} 0 & M \\ M & C \end{pmatrix}\dot{y}\ +\ \begin{pmatrix} -M & 0 \\ 0 & K \end{pmatrix}y\ =\ \begin{pmatrix} 0 \\ \ell\,(t) \end{pmatrix} \qquad (2.12)$$

This form of the Duncan transformation, and its corresponding eigenvalue problem, have been widely used in structural dynamics. It is occasionally advantageous (see Brandon (1984a)), as will be described, to use an alternative form of the Duncan formulation based on the identity

$$K\,\dot{x}\ -\ K\,\dot{x}\ =\ 0$$

with y defined as before, to give

$$\begin{pmatrix} M & 0 \\ 0 & -K \end{pmatrix}\dot{y}\ +\ \begin{pmatrix} C & K \\ K & 0 \end{pmatrix}y\ =\ \begin{pmatrix} \ell(t) \\ 0 \end{pmatrix} \qquad (2.13)$$

At first sight it would appear that the eigenvalue problem corresponding to this equation (or equation (2.12)) is no more arduous than that of equation(2.5). This is however far from the truth. The Duncan form, and indeed any method involving general distributions of viscous damping, requires the use of complex arithmetic (for both eigenvalues and eigenvectors). In addition the advantageous numerical properties of the undamped case can no longer be guaranteed. Indeed the existence of a full set of eigenvectors for general eigenvalue problems cannot always be demonstrated.

2.5 ORTHOGONALITY PROPERTIES OF THE EIGENVALUE PROBLEM

A central feature of the solution and use of eigenvalue problems is the orthogonality property of the eigenvectors, leading to the reforming of sets of linear equations in terms of the spectral decomposition of the system. Consider initially the standard symmetric eigenvalue problem defined above

$$(X - \mu I) \xi = 0 \qquad (2.14)$$

taking two simple (ie not repeated) eigenvalues μ_r, μ_s and their corresponding eigenvectors ξ_r and ξ_s respectively. Substituting these values in (2.14) gives

$$X \xi_r - \mu_r \xi_r = 0 \qquad (2.15)$$

$$X \xi_s - \mu_s \xi_s = 0 \qquad (2.16)$$

Premultiplying (2.15) by ξ_s^t and (2.16) by ξ_r^t and subtracting gives

$$(\mu_r - \mu_s) \xi_r^t \xi_s = 0 \qquad (2.17)$$

Since it has been stipulated that the eigenvalues μ_r and μ_s are distinct, it must be concluded that the vector inner product $\xi_s^t \xi_r$ is zero, ie the eigenvectors are orthogonal. By back-substitution it can also be seen that $\xi_s^t \, X \, \xi_r$ is also zero.

Considering now the generalised eigenvalue problem for the conservative system, equation (2.5), a similar orthogonality condition can be identified

$$x_r^t \, M \, x_s \, = \, \begin{cases} m_r & r = s \\ 0 & r \neq s \end{cases}$$

$$x_r^t \, K \, x_s \, = \, \begin{cases} k_r = \lambda_r^2 \, m_r & r = s \\ 0 & r \neq s \end{cases}$$

Ordering the eigenvalues such that $\lambda_1^2 \leq \lambda_2^2 \leq \, ..\leq \lambda_n^2$ the corresponding eigenvectors may be used to construct the modal matrix

$$\Phi \, = \, \left\{ \, \phi_1 \ \phi_2 \ \phi_3 \, \ldots \, \phi_n \, \right\}$$

The orthogonality conditions may now be expressed in matrix form

$$\Phi^t \, M \, \Phi \, = \, \text{diag} \left\{ m_i \right\} \, = \, m \qquad\qquad (2.18)$$

$$\Phi^t \, K \, \Phi \, = \, \text{diag} \left\{ k_i \right\} \, = \, k \qquad\qquad (2.19)$$

These matrices are often referred as the modal mass and modal stiffness matrices, although their dimensions are energy/(time squared) and energy repectively.

It frequently occurs that experimental modal analysis provides estimates of modal masses and modal stiffnesses. In theory (but not in practice) the system mass and stiffness matrices may be constructed

$$M = \Phi^{-t} m \Phi^{-1} \qquad (2.20)$$

$$K = \Phi^{-t} k \Phi^{-1} \qquad (2.21)$$

The limitations of the practical application of these equations will be discussed further, later in the text.

In conservative problems, ie excluding damping, it can be shown that a complete set of orthogonal eigenvectors can always be constructed even for the case of equal eigenvalues (see Bishop, Gladwell and Michaelson (1965)). This is not true generally for the dissipative (damped) case as the justification relies on the positive definiteness of the system matrices, which is no longer true in the case of equations (2.12) and (2.13).

2.6 CONSTRUCTING RECEPTANCE REPRESENTATIONS

The modal representation, in terms of the eigenvalues and eigenvectors, is not particularly useful in isolation. The structural properties most often required are the force-response characterisics of the structure. In the current work the emphasis will be on the force-displacement transfer function, the receptance. Other common tranfer functions based on force-velocity and force-acceleration are discussed by Ewins (1984). Taking the Laplace transform of equation (2.3) and neglecting initial conditions gives

$$S \mathfrak{X} = (s^2 M + s C + K) \mathfrak{X} = \mathfrak{f}(s) \qquad (2.22)$$

The matrix S, equal to the term in brackets, is referred to as the dynamic stiffness matrix. Provided that this term is non-singular then $\mathfrak{X}$ is given by

$$\mathscr{X} = (s^2 M + s C + K)^{-1} \mathscr{\ell} (s) \qquad (2.23)$$

The condition for non-singularity is that $s \neq \lambda_i$, ie the solution cannot be evaluated when the structure is excited at a resonance. The matrix $R = S^{-1} = (s^2 M + s C + K)^{-1}$ is known as the receptance matrix. The inversion implied in equation (2.23) is not practicable for realistic engineering structures and alternative methods have been derived to generate a sufficiently accurate approximation to the receptance matrix for practical purposes. The most common form is the spectral decomposition of the receptance matrix, in terms of eigenvectors and modal masses. Considering initially the undamped system, then R is given by

$$R_c = (s^2 M + K)^{-1} \qquad (2.24)$$

where the subscript c denotes the equivalent conservative system. Using the modal decomposition of the mass and stiffness matrices, equations (2.20) and (2.21) respectively, gives

$$R_c = \left\{ \Phi^{-t} \left\{ s^2 m + k \right\} \Phi^{-1} \right\}^{-1} \qquad (2.25)$$

The inversion of this matrix is extremely straightforward, since the modal matrix is known and the matrix ($s^2 m + k$) is diagonal, with a trivial inversion giving

$$R_c = \Phi \ diag \left\{ \frac{1}{s^2 m_i + k_i} \right\} \Phi^t \qquad (2.26)$$

This is more commonly expressed as the summation of a

series of vector products

$$R_c = \sum_{i=1}^{n} \frac{\phi_i \, \phi_i^t}{s^2 m_i + k_i} \qquad (2.27)$$

In many practical applications it is not practicable, and fortunately not necessary, to evaluate all terms in this series, since many will be negligible within the frequency range of interest. The truncation of the series is often described as an incomplete model, a term due to Berman and Flannelly (1971). The analysis of incomplete models, and their analytical consequences, will be of central importance in the subsequent discussion.

2.7 PROPORTIONAL DAMPING

An intermediate stage between undamped systems and general damping distributions was originally given in concept by Rayleigh (1945), although expression in matrix form was provided by Foss (1956) and subsequently developed further by Caughey (1960) and Caughey and O'Kelly (1965). Proportional damping (sometimes also referred as classical damping) relies on the damping matrix having a structure compatible with the mass and stiffness matrices. The receptance matrix for the damped case is not generally expressible in the same spectral form of equation (2.27) because there is no guarantee that the modal matrix transformation, which simultaneously diagonalises the mass and stiffness matrices, will also diagonalise the damping matrix. Hence the modal dynamic stiffness matrix for the damped case is not necessarily diagonal and its inversion is no longer trivial. The basis of proportional or classical damping is the identification of damping distributions which are diagonalised by the modal transformation defined above. The earliest ideas involved a simple multiple of either the mass or stiffness matrix (hence proportional damping). Obviously if the damping matrix is a simple multiple of either the mass or stiffness matrix then it will be diagonalised by the same modal transformation. Subsequent work by Foss (1956), Caughey (1960) and Caughey and O'Kelly (1965) addressed the problem of sufficient conditions for a damping matrix to be diagonalised by the same modal transformation.

Although the original model for proportional damping (ie a simple multiple of mass or stiffness) may seem crude, there are many structures where it is a satisfactory approximation, for example light hysteretic damping in monolithic structures. It was shown by Caughey (1960) that a sufficient condition for simultaneous diagonalisation is that C is given by

$$M^{-1} C = \sum_{i=1}^{\infty} a_i \left\{ M^{-1} K \right\}^i$$

Classical damping is therefore characterised by the equation

$$\Phi^t C \Phi = \text{diag} \left\{ c_i \right\} = c \qquad (2.28)$$

The consequences for the solution of the eigenvalue problem are that, although the eigenvalues of the quadratic eigenvalue problem now require complex arithmetic for their computation, the eigenvectors of the quadratic problem are the same, and consequently real, as those of the allied conservative problem. The receptance for the classically damped system may be constructed using the same spectral form as for the undamped case

$$R_d = \sum_{i=1}^{n} \left\{ \frac{\Phi_i \Phi_i^t}{s^2 m_i + s c_i + k_i} \right\} \qquad (2.29)$$

2.8 PHYSICAL SIGNIFICANCE OF COMPLEX REPRESENTATION

The physical significance of the eigenvalues is identical to that in single degree of freedom systems: the real part of the solution will be negative, for passive systems, and represents the dissipation; the imaginary part represents the angular frequency in the harmonic component (see for example Meirovitch (1986)). The significance of a complex eigenvector is less well known and is occasionally misunderstood or misrepresented in the literature. As has been described above, in the analysis of hysteretic damping, the differences in argument of the complex numbers correspond to a phase shift (or time delay). Thus, for a structure with a general damping distribution, different points on the structure will have varying phase relationships with respect to the input force, giving the effect of a travelling wave. In contrast, for proportionally damped structures all points on the structure will be in phase with each other but not necessarily in phase with the driving force. This effect is consistent with the behaviour of a standing wave. The assumption of proportional damping underlies one of the most highly regarded experimental methods for resonance testing, that of Traill-Nash (1958) (for an appraisal of early methods of testing in structural dynamics see Bishop and Johnson (1963)). In this technique, which pre-dates methods which rely on the Fast Fourier Transform, an array of shakers simultaneously excite the structure sinusoidally at a frequency identified as a natural frequency by an initial frequency search. Originally based on analogue instrumentation, the method has subsequently been used using digital control analysis hardware (see for example Gabri and Matthews (1980) and Gage (1986)).

2.9 RECEPTANCE MATRICES FOR GENERAL DAMPING DISTRIBUTIONS

The receptance matrix for structures with general damping distributions can be constructed from the Duncan formulation, equation (2.12).

$$
\begin{pmatrix} 0 & M \\ M & C \end{pmatrix} \dot{y} + \begin{pmatrix} -M & 0 \\ 0 & K \end{pmatrix} y = \begin{pmatrix} 0 \\ \ell(t) \end{pmatrix} \quad (2.12)
$$

As before the receptance matrix may be constructed using the spectral decomposition of the solution to the eigenvalue problem. Defining matrices A and B

$$
A = \begin{pmatrix} 0 & M \\ M & C \end{pmatrix} \qquad B = \begin{pmatrix} -M & 0 \\ 0 & K \end{pmatrix}
$$

and 2n eigenvectors ψ_i and eigenvalues λ_i, allows the diagonalisation of A and B (subject to the rather pedantic proviso that the generalised eigenvalue problem has a full system of eigenvectors). These eigenvectors and eigenvalues are either real or occur in complex conjugate pairs. As before, the eigenvectors diagonalise the system matrices

$$
a_i = \psi_i^t A \psi_i \qquad\qquad b_i = \psi_i^t B \psi_i
$$

Considering the construction of ψ_i

$$
\psi_i = \begin{pmatrix} \lambda_i \phi_i \\ \phi_i \end{pmatrix}
$$

leads to the following expressions for a_i and b_i

$$
a_i = 2 \lambda_i \phi_i^t M \phi_i + \phi_i^t C \phi_i
$$

$$
b_i = - \lambda_i^2 \phi_i^t M \phi_i + \phi_i^t K \phi_i
$$

giving the state vector y in the form

$$y = \sum_{i=1}^{2n} \frac{\psi_i \ \psi_i^t}{s \ a_i + b_i} \begin{pmatrix} 0 \\ \ell(s) \end{pmatrix} \qquad (2.30)$$

The receptance relates the lower half of y to the force vector $\ell(s)$, which may be achieved by an appropriate partitioning of the vectors

$$x = \sum_{i=1}^{2n} \frac{\phi_i \ \phi_i^t}{s \ a_i + b_i} \ \ell(s) \qquad (2.31)$$

2.10 ORDER OF SYSTEM MODELS

As was suggested in section 2.1 the discrete model is an (often acceptable) approximation to the real structure. The order of a model (corresponding to the dimension of the system matrices), particularly under experimental conditions, often depends on factors beyond the control of the analyst.

This may be constrained by the number of channels of data available on the instrumentation (continually increasing) or by access limitations to the structure under test. In contrast, finite element models usually have a large number of degrees of freedom but often suffer from inadequacies due both to necessary geometrical simplification of the model and also to the neglect of damping. It is necessary therefore, in the study of reanalysis techniques, to consider the differences between the commonly used models and the methods of reconciliation of the information provided by analysis of these structural representations.

2.11 BRACKETING PROPERTIES OF EIGENVALUES

The eigenvalues for the generalised eigenvalue problem given by equation (2.5) (λ^2 M + K) ϕ = 0 are often expressed in terms of the Rayleigh quotient

$$\lambda^2_i = \frac{\phi^t_i \ K \ \phi_i}{\phi^t_i \ M \ \phi_i}$$

(2.32)

It can readily be seen that an increase in stiffness will give rise to an increase in the magnitude of the eigenvalue, whereas an increase in mass will cause a decrease in the eigenvalue. The sole exception is the case when the modification is located at a node of the mode, when the eigenvalue will be unchanged. The change in the value of the Rayleigh quotient is however not unlimited, as it is actually asymptotic, a fact exploited by Weissenburger (1966,1968), Pomazal (1969) and Pomazal and Snyder (1970). This will be discussed in Chapter 3. The bracketing or inclusion theorems required to understand that theory are also of importance in the comparison of alternative models of the same structure, particularly the reconciliation of experimental and finite element models. The description given here is primarily qualitative since the detail of the theory is well established, see for example Meirovitch (1967, 1980, 1986). The bracketing (or inclusion) theorems relate the eigenvalues of structural approximations when the number of degrees of freedom is varied. Consider the eigenvalues of equation (2.5) computed according to equation (2.32). If the number of degrees of freedom is reduced by one this is equivalent to imposing an additional constraint on the system, making it stiffer and raising the eigenvalues of the system. If these new eigenvalues are denoted by λ^2_{ic} (c for constrained) then the inclusion theorem (Meirovitch (1986) p278) gives

$$\lambda_1^2 \;\le\; \lambda_{1c}^2 \;\le\; \lambda_2^2 \;\le\; \lambda_{2c}^2 \ldots\ldots \;\; \lambda_{(n-1)c}^2 \;\le\; \lambda_n^2$$

$$(2.33)$$

Thus each eigenvalue of the reduced system is bounded or "bracketed" by the pair of eigenvalues of the initial system. Conversely, increasing the order of model has the effect of reducing the estimates of the natural frequencies. The logical extension of this argument is that the continuum model of the structure, viewed as the limit of a sequence of discrete approximations, will give an absolute lower bound to the estimates of any discrete model, and the larger the number of degrees of freedom, the better will be the approximation.

2.12 ANALYTICAL METHODS OF ORDER REDUCTION

Repeated application of the inclusion theorem stated above leads to an unduly pessimistic view of the quality of low order structural models. For example the introduction of (say) four degrees of freedom into a model may reduce the estimate of the fourth eigenvalue of the augmented model below the first eigenvalue of the initial approximation.

In practice, however, the numerical properties of the Rayleigh quotient usually lead to excellent estimates of the lower eigenvalues, even for rather crude approximations to the eigenvector and coarse spatial discretisation. The quality of the estimation for low order models, both of eigenvalues and eigenvectors, deteriorates progressively as higher values of eigenvalue are taken.

The most common technique for order reduction in analytical structural dynamics is widely credited to Guyan (1965) (and hence called Guyan reduction) although a similar technique had been proposed by Irons (1963) previously (see also Irons (1965)).

In finite element work, degrees of freedom are partitioned into "master" and "slave" nodes, the slave nodes being eliminated according to some pre-set scheme during assembly of the system matrices, a necessary condition being that slave nodes must not be points of application of external forces. From the initial proposals of Guyan and Irons there has been a steady progression of development of new methods which improve the closeness of original and reduced systems (see Henshell and Ong (1975), Miller (1980), Shah and Raymund (1982) and Paz (1984)).

Concurrent with the use of order reduction techniques in finite element analysis has been the development of substructuring techniques (see Hale and Warren (1983) for a representative bibliography). It is common in such applications to solve substructure dynamic equations without performing order reduction (often referred to as condensation) and achieve the required reduction in the order of the problem by deleting substructure modes higher than a certain frequency range prior to synthesis of the combined system.

A conservative system may be partitioned into master and slave nodes:

$$\left\{ s^2 \begin{pmatrix} M_{mm} & M_{ms} \\ M_{sm} & M_{ss} \end{pmatrix} + \begin{pmatrix} K_{mm} & K_{ms} \\ K_{sm} & K_{ss} \end{pmatrix} \right\} \begin{pmatrix} æ_m \\ æ_s \end{pmatrix} = \begin{pmatrix} \ell(s) \\ 0 \end{pmatrix}$$

(2.34)

or, using the dynamic stiffness matrix

$$\begin{pmatrix} S_{mm} & S_{ms} \\ S_{sm} & S_{ss} \end{pmatrix} \begin{pmatrix} æ_m \\ æ_s \end{pmatrix} = \begin{pmatrix} \ell(s) \\ 0 \end{pmatrix}$$

(2.35)

considering the lower half of the partition

$$S_{sm} \, æ_m \; + \; S_{ss} \, æ_s \quad = \quad 0$$

the slave coordinates $æ_s$ can be expressed in terms of master coordinates $æ_m$

$$æ_s \; = \; - \, S_{ss}^{-1} \, S_{sm} \, æ_m \qquad\qquad (2.36)$$

subject to the condition $| \, S_{ss} \, | \neq 0$, giving a reduced problem

$$\left(S_{mm} \; - \; S_{ms} \, S_{ss}^{-1} \, S_{sm} \right) \, æ_m \; = \; \ell(s) \qquad (2.37)$$

or more fully

$$\left\{ \left(s^2 \, M_{mm} + K_{mm} \right) \; + \right.$$

$$\left. \left(s^2 \, M_{ms} + K_{ms} \right) \left(s^2 \, M_{ss} + K_{ss} \right)^{-1} \left(s^2 \, M_{sm} + K_{sm} \right) \right\} \, æ_m \; = \; \ell(s)$$

$$(2.38)$$

This is a special case of the inverse operation to the formation of the Duncan transformation described in section 2.4. A discussion of the application of (2.38) has been given by Leung (1978,1979).

The Guyan reduction, often also described as static condensation, discounts the frequency dependency of the coordinate transformation of equation (2.36). The analytical consequences are discussed by Fox (1981) (see also Utku, Clemente and Salama (1985)). It should be noted that, latterly, Irons (1981) expressed considerable disquiet concerning the use of Eigenvalue Economisers:

".. the writer remains unenthusiastic, despite the good results, and regrets his earlier association with eigenvalue economisers... Unskilled people use commercial packages; such algorithms as these, which usually give good answers but can behave very badly, are simply not suitable".

A dynamic condensation (ie applying (2.36) exactly) is often assumed implicitly in experimental work. Because of the difficulties of measuring rotational deformations using standard vibration testing hardware (see for example Ewins (1984) pp146-8) it is not unusual to assume that a meaningful translational mass matrix and stiffness matrix are an adequate representation the structure. The structural identification then proceeds on this basis. This may be by implication only and it is not always clear that the analyst is aware of the criteria for acceptability. The assumption is equivalent to assigning the rotational components to the slave coordinates. Two necessary conditions for the validity of such a model are:

(i) $\left| s^2 M_{ss} + K_{ss} \right| \neq 0$

for all s within the bandwidth of experimental measurement (ie the rotational subsystem must have no resonances within the chosen frequency band);

(ii) The coupling matrices M_{sm} and K_{sm} (and hence their transposes M_{ms} and K_{ms} respectively) must be small in comparison to the direct matrices M_{mm}, M_{ss}, K_{mm} and K_{ss}. This would be measured in a suitable matrix norm, eg the Frobenius norm (also called the

Euclidean or Schurr norm) $\left\| A \right\|_F = \sqrt{\sum_{i=1}^{n} \sum_{i=1}^{n} (a_{ij})^2}$.

If these two conditions are met then the translational matrices M_{mm} and K_{mm} will be satisfactory representations of the identified system since the remaining terms will be small by comparison. If however these conditions do not apply then there are potentially serious consequences in assuming a purely translational model. This will be particularly important in experimental system identification where a completely inadequate representation of the energy distribution of the system would be made.

2.13 SYNTHESIS OF STRUCTURAL MODELS FROM EXPERIMENTAL DATA: INCOMPLETE MODELS

It was suggested, in section 2.5, that if all of the modes (eigenvectors) and the modal masses m_i and stiffnesses k_i are known then the system mass and stiffness matrices may be constructed using equations (2.20) and (2.21):

$$M = \Phi^{-t} \, m \, \Phi^{-1} \tag{2.20}$$

$$K = \Phi^{-t} \, k \, \Phi^{-1} \tag{2.21}$$

The identification of the complete system matrices of a large discrete system of order n is not possible in practice for a number of reasons:

(i) the instrumentation is restricted to a limited frequency range and will be incapable of identifying high frequency modes;

(ii) the necessary inversions will be subject to the problems of numerical ill-conditioning (see Flannelly and Berman (1972));

(iii) the discrepancy between discrete and continuum models increases with increasing frequency.

Recalling the construction of the receptance matrix in spectral form (equations (2.26),(2.29) and (2.31),

$$R_c = \Phi \, diag \left\{ \frac{1}{s^2 \, m_i + k_i} \right\} \Phi^t \qquad (2.26)$$

$$R_d = \sum_{i=1}^{n} \left\{ \frac{\phi_i \, \phi_i^t}{s^2 \, m_i + s \, c_i + k_i} \right\} \qquad (2.29)$$

$$\mathcal{R} = \sum_{i=1}^{2n} \frac{\phi_i \, \phi_i^t}{s \, a_i + b_i} \, \ell \, (s) \qquad (2.31)$$

it can be seen that the response of the system can be considered as the superposition of the contributions of the responses of the individual modes (the fundamental characteristic of linear systems). Furthermore the response due to individual modes is a dominant feature of the overall response in the region of its natural frequency. This property is the basis for single mode curve fitting algorithms (see Bishop and Johnson (1963), for the basic theory, and Brandon and Cowley (1983) for a modern implementation). The receptance matrix is often closely approximated by the use of two residual terms to represent out-of-bandwidth response properties, ie

$$R = \frac{1}{s^2} \, R_m + \sum_{i=1}^{u} \left(\frac{\phi_i \, \phi_i^t}{s^2 \, m_i + k_i} \right) + R_k \qquad (2.39)$$

where the upper and lower residual matrices, R_m and R_k respectively, are constant and the summation now only applies over the measured modes. The former corresponds to an unconstrained mass effect, or rigid body mode, whilst the latter corresponds to a statically compliant constraint distribution. Much of the relevant theory was developed by van Loon (1974) and Klosterman (1971). This type of response model is well proven and widely used (see for example Ewins (1980)).

More controversial, yet as justified on theoretical grounds, has been the construction of approximations to the mass and stiffness matrices of the structure using the measured modal data using rank deficient equivalents to equations (2.20) and (2.21):

$$M_r = \Phi_r^{gt} \, m_r \, \Phi_r^g \qquad\qquad (2.40)$$

$$K_r = \Phi_r^{gt} \, k_r \, \Phi_r^g \qquad\qquad (2.41)$$

where Φ_r is now nxr, and Φ_r^g denotes the Moore-Penrose generalised inverse of Φ_r (often also called the pseudo-inverse). The definition and application of the pseudo-inverse has been the subject of some misunderstanding and abuse in structural dynamics (see Brandon (1988)). The derivation and properties are discussed in Appendix 1. Although the matrices M_r and K_r are of rank r, ie they have r linearly independent rows/columns, they are of order n. The essential property of such a construction for approximations to the mass and stiffness matrices is that these matrices define an eigenvalue problem which replicates the measured properties of the structure. The term "incomplete models" was used by Berman and Flannelly (1971) to describe mathematical models depending on this type of representation. Such representations are widely mistrusted within the modal analysis community because of gross discrepancies between the incomplete mass and stiffness matrices, produced from experimental tests, and the corresponding analytically constructed full rank matrices of the same order. Thus the lack of physical significance which can be derived from the models is a bar to their acceptance. The idea of an incomplete model was originally rationalised by Ross (1971) in the following terms:

"For a forced response analysis, the individual matrix elements need not have physical significance because their primary function is to model accurately the structure's dynamic response. However, when synthesizing high-order models which are to be compared with high-order theoretical models, it is necessary that the individual elements of the synthesized matrices agree with the well-defined meanings of the theoretic elements."

A similar approach to that of Ross was initially made by Berman (1975), although he and his associates have latterly used these methods primarily for adjustment to an existing model (Berman, Wei and Rao (1980), Berman (1980), Berman and Nagy (1983)). The basic method follows the work of Collins, Young and Kiefling (1972). Other significant contributions have been due to Chen and Garba (1979), Baruch and Bar Ishtak (1978), Baruch (1984) and Natke and Schulze (1981). Latterly work has been published by Dobson et al (1984).

In evaluating these methods the viewpoint, and hence the objectives, of the analyst must be considered. Although it is still an active area of research, the attempts to construct physically meaningful mass and stiffness matrices from experimental data will inevitably be defeated by the numerical characteristics of the process, described at the beginning of this section. This fatuous practice was subjected to an articulate critique by Berman (1984), and will not be pursued further here. This type of method has shown considerable promise for adjustment of established analytical models and the author would have few reservations in applying such techniques to parameter adjustment. There is little doubt that these procedures are less prone to (possibly catastrophic) errors than those which will be described in the remainder of this chapter.

2.14 CONSTRUCTING ANALYTICAL MODELS FROM EXPERIMENTAL DATA: LOW ORDER MODELS

The methods described in the previous section involved low rank (the number modes, r, used in model construction) but large order (the number of degrees of freedom considered, n). The rank of the matrix indicates the number of independent vectors used in the construction of the rows (or columns) of the matrix.(The rank of a system estimated from experimental testing can be an extremely subjective measure- see Brandon (1987).)

In identifying the system in terms of mode shapes and frequency response functions (corresponding to the information given by the eigenvalues of the analytical system), it may be sufficient (depending on the requirements of the analyst), to undertake the modal survey at only r points (subject to the pedantic constraint that the r points are not all nodes of at least one mode of the structure, in which case that mode, or modes,will not be detected and the system will appear rank deficient). In this case the mass and stiffness matrices defined by equations(2.20) and (2.21) will define an eigenvalue problem which will reproduce exactly the observed response behaviour of the structure at the chosen points, although this model may be extremely sensitive to small perturbations in the structural parameters.

This type of representation is referred to as a low order model of the structure. The practice of representing distributed structural properties by local concentrations is often described as "lumping" and the resulting models lumped parameter models. The analysis of such models has progressed from the initial ideas of Ross (1971). Successful application of such methods has been reported by Lincoln (1977), Tlusty and Morikawa (1976) and Tlusty, Ismail and Prossler (1978).

Whittaker and Sadek (1980) have proposed the incorporation of low order models in

".. a complete on-line optimisation package with regard to the dynamics of machine tools."

The successful application of low order models has been achieved almost exclusively when applied to machine tool structures. There is, however, widespread scepticism about these methods and anecdotal evidence that the methods break down when applied to more general structures. (Little has been published since many researchers are understandably reluctant to report their failures). That is not to say however that the machine tool researchers have been misled, since machine tools are intrinsically amenable to lumped parameter representations, comprising rigid massive structural members, such as beds, columns, etc, connected by light compliant interfaces at guideways, slides etc. Further, the structural characteristics of interest, the subject of the type of optimisation described by Whittaker and Sadek (1980), are often restricted to the most compliant mode, (measured at the tool-workpiece interface), which Rayleigh's principle suggests will be closely approximated with a low order model.

The low order model may be satisfactory for response calculations but it has serious deficiencies for reanalysis. This stems from the basic mismatch between the model and the observed data. The experimental data are discrete observations of the continuous structure implying, from the inclusion principle, that the natural frequencies measured will provide absolute lower bounds to the frequencies predicted from a discrete representation. Conversely, if estimates of the structure's natural frequencies are provided analytically, through a process of order reduction to match the degrees of freedom of the low order model, then the analytically derived frequencies will be (perhaps substantially) greater than those measured. Thus for reanalysis purposes, although the measured mode shapes may conform closely to those which would be given by an analytical low order model, the measured frequencies are likely to be substantial underestimates, becoming progressively worse with increasing frequency.

2.15 INCREASING THE ORDER OF MODELS: THE ESCALATOR METHOD

Of considerable historical importance, though rarely studied latterly, is the escalator method. From an analytical viewpoint this method is the inverse of the exact dynamic order reduction methods described by Leung (1978,1979). Although now of little practical importance, the method was used by Weissenburger (1966,1968) to develop exact modal reanalysis techniques, described in Chapter 3. The method originates with the difficulties involved in solving large scale eigenvalue problems with very limited computational facilities. An initial solution to a 2x2 problem is used to solve a 3x3 problem by adjoining a border to the matrix (a "gnomon"). This solution is then used to solve a bordered 4x4 system. The system is then successively "escalated" until the desired order is attained, with each solution developed from the preceding level. Much of the development of the method is due to Morris (1947), although the contributions of Fox (1952) are also significant.

CHAPTER 3
Exact Modal Methods

3.1 EXACT SOLUTIONS FOR THE EIGENVALUE PROBLEM: SCOPE AND APPLICABILITY

Suppose that there is available a complete solution set for the eigenvalue problem

$$(\lambda^2 M + K) \phi = 0 \qquad (3.1)$$

or

$$(\lambda^2 M + \lambda C + K) \phi = 0 \qquad (3.2)$$

where M, as before, is nxn. It may be desired either to evaluate the effect of a localized modification or to compare the effect of several such modifications. Since the cost of repeated solution of the full eigenvalue problem may be prohibitively expensive, it is necessary to consider whether the variant problem(s) may be solved using the previously computed solution.

The methods which are described in this chapter involve an analytically exact formula, which combines the modal properties (eigenvalues and eigenvectors) of the unmodified structure with a modification which is (analytically at least) unlimited in magnitude but restricted in distribution, to give the modal properties of the modified structure. In the methods described, this exact problem is, however, solved using an iterative algorithm. This was originally derived for conservative structures by Weissenburger (1966, 1968) and extended to viscously damped systems by Pomazal (1969) and Pomazal and Snyder (1970) (although it should be noted that Weissenburger had given an outline analysis for general eigensystems in an appendix to his DSc Dissertation (1966)). The method given by Pomazal was primarily suitable for stiffness and damping modifications, the incorporation of mass modifications leading to a significantly more complicated procedure. Extension to efficient incorporation of mass modifications was devised initially by Hallquist (1976), although this formulation entailed sacrificing the symmetry of the eigenvalue problem. A method for efficient mass modification, which retained the symmetry of the eigenvalue problem, was subsequently provided by Brandon (1984c). The problem has recently been addressed independently by Skingle and Ewins (1989), based largely on an experimental interpretation of the structural model.

3.2 MODIFICATION ANALYSIS FOR CONSERVATIVE SYSTEMS

The theory developed by Weissenburger (1966, 1968) is applicable to continuous structures (and was derived by Weissenburger using a Lagrangian formulation resulting in modal equations based on energy functionals rather than the inner products commonly used in discrete systems). The discrete mass modification is introduced by the use of the Dirac delta function. For both discrete and continuous systems the coordinate transformation to principal (or modal) coordinates is accomplished by forming energy inner products as in equations (2.18) and (2.19), although in the continuous case the order of the resulting matrix is infinite (see Lanczos(1957)).

3.2.1 Effect of Concentrated Mass or "Earthed" Stiffness

Consider initially a mass modification, represented by an increment in the rth coordinate, typical of the attachment of a "lumped mass". This modification results in the simplest incremental matrix comprising all zeroes except for the single lumped mass term in the (r,r) position of the matrix. A similar modification to either the damping or stiffness matrix implies a compliant connection to "earth".

Symbolically the modified system is represented by the equation

$$(M + \delta m \, e_r \, e_r^t) \, \ddot{x} + K \, x = f(t) \tag{3.3}$$

where e_r is the column vector with 1 in the r position and zero elsewhere (Thus $e_r \, e_r^t$ is a matrix with 1 in the r-th diagonal position and zero elsewhere).

As has been shown in section 2.6, the modal decomposition of the mass and stiffness matrices, given by equations (2.18) and (2.19), allows the response problem for the unmodified system to be transformed to a superposition of the effects of the responses of the individual modes, ie the response of the coupled n degree of freedom system is resolved into the simpler problem of the summation of the responses of n separate single degree of freedom systems. This is often demonstrated using the transformation to principal or modal coordinates. Defining q

$$q = \Phi^{-1} \, x \tag{3.4}$$

ie

$$x = \Phi \, q \tag{3.5}$$

then this transformation "uncouples" the response

equations (ie converts the nxn response problem to the n individual modal contributions- the spectral decomposition). This transformation may be used to construct the spectral decomposition of the receptance matrix derived in section 2.6 (cf equation (2.27)).

Consider, for example, the excitation of the undamped system in Laplace form (the undamped equivalent of equation (2.22))

$$(s^2 M + K) \, \mathit{x} \; = \; \mathit{f}(s) \tag{3.6}$$

Applying the coordinate transformation to modal coordinates (equation (3.5)) gives

$$(s^2 M \Phi \; + \; K \Phi) \, q = \mathit{f}(s) \tag{3.7}$$

Premultiplying by Φ^t gives

$$(s^2 \Phi^t M \Phi + \Phi^t K \Phi) \, q = \Phi^t \, \mathit{f}(s)$$

Substituting equations (2.18) and (2.19)

$$(s^2 m \; + \; k \;) \, q \; = \; \Phi^t \, \mathit{f}(s) \tag{3.8}$$

Since both m and k are diagonal matrices this provides the desired uncoupling, ie the replacement of an n degree of freedom matrix system of equations with n independent single degree of freedom systems, which may be treated as scalar equations.

If, however, this transformation is applied to the modified system (equation (3.3)) then, although the coordinate transformation to modal coordinates diagonalizes the mass and stiffness matrices, as before, the modification matrix, instead of only containing one non-zero term, becomes full (although still of rank one).

The Laplace form of equation (3.3) becomes

$$(s^2(m + \delta m\, r\, r^t) + k)\, q = \Phi^t \ell(s) \qquad (3.9)$$

where $r = \Phi^t\, e_r$, the rth column of Φ^t.

Now, instead of solving the system by the trivial inversion of a diagonal matrix, the dynamic stiffness matrix, in the first set of modal coordinates, is once again full. As before, the solution to the system can be achieved by solving the new eigenvalue problem

$$(\sigma^2(m + \delta m\, r\, r^t) + k)\, \varphi = 0 \qquad (3.10)$$

and combining the solution with the modal solution to the unmodified problem. (The two modal matrices are multiplied together to transform to the initial spatial coordinates).

Thus the modification to the system results in a new eigenvalue problem of equal complexity to the original problem. Apparently therefore nothing is gained over re-solving the new system.

Weissenburger however recognized that the structure of the modification matrix, specifically its unity rank, allowed the expression of the eigenvalues and eigenvectors of the modified system to be derived recursively from those of the unmodified system.

The exploitation of low rank of the modification matrix is not exclusive to Weissenburger's method. The method employed by Mahalingam (1975a,b), often described as the pseudoforce method, uses a similar strategy. It has been demonstrated by Brandon et al(1988) that the pseudoforce method is equivalent to Kron's method (1963).

Recalling the relationship between the non-zero diagonal elements of the modal mass and modal stiffness matrices (from section (2.5))

$$k_i = -\lambda_i^2\, m_i$$

equation (3.10) can be rearranged

$$\left(\text{diag} \left(s^2 - \lambda_i^2 \right) m \right) \varphi = - \delta m \, s^2 \, r \, r^t \, \varphi \qquad (3.11)$$

The effect of the diagonal matrix on the left hand side of equation (3.11) is that the coefficients of the eigenvector of the modified system are re-scaled to reflect the weightings of the modal masses and stiffnesses of the modes of the original system. It is possible therefore to isolate the effect of the modification due to each mode in turn.

The constraint conditions imposed by the unit rank of the modification matrix allow each row of equation (3.11) to be computed individually to give

$$\left(s^2 - \lambda_i^2 \right) m_i \, \varphi^{(i)} = - \delta m \, s^2 \, r_i \sum_{k=1}^{n} r_k \, \varphi^{(k)} \qquad (3.12)$$

where $\varphi^{(i)}$ denotes the i-th element of the vector φ. Re-arranging (3.12) gives

$$\sum_{k=1}^{n} r_k \, \varphi^{(k)} = \frac{\left(s^2 - \lambda_i^2 \right) m_i \, \varphi^{(i)}}{- \delta m \, s^2 \, r_i} \qquad (3.13)$$

Extending the analysis for the other rows will result in similar expressions, with the left hand side of (3.13) being common to each, ie

$$\frac{\left(s^2 - \lambda_1^2 \right) m_1 \, \varphi^{(1)}}{- \delta m \, s^2 \, r_1} = \frac{\left(s^2 - \lambda_2^2 \right) m_2 \, \varphi^{(2)}}{- \delta m \, s^2 \, r_2} = \dots \qquad (3.14)$$

These relationships may be substituted back into equation (3.12)

$$\frac{1}{\delta m \ s^2} + \sum \frac{r_i}{m_i \left(s^2 - \lambda_i^2 \right)} = 0$$

(3.15)

which is the characteristic equation of the eigenvalue problem (equation (3.10)). Since the eigenvector is only determined to a multiplicative constant, equation (3.14) gives the eigenvector

$$\varphi = \beta \left(\frac{r_1}{m_1 \left(s^2 - \lambda_1^2 \right)} \quad \frac{r_2}{m_2 \left(s^2 - \lambda_2^2 \right)} \cdot \cdot \frac{r_n}{m_n \left(s^2 - \lambda_n^2 \right)} \right)^t$$

(3.16)

The solution to equation (3.15) is obtained by considering the first and second terms of the equation separately.

Consider first the form of a single term of the summation in equation (3.15) as s varies. This is shown in figure 6. The curve is of hyperbolic form with the original eigenvalue forming an asymptote.

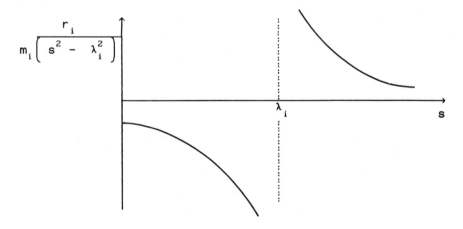

FIGURE 6: EFFECT OF SINGLE TERM OF AUXILIARY FUNCTION

48

The effect of the summation of the contributions from all the modes leads to the curve shown in figure 7. The curve is bounded asymptotically by the eigenvalues of the initial system.

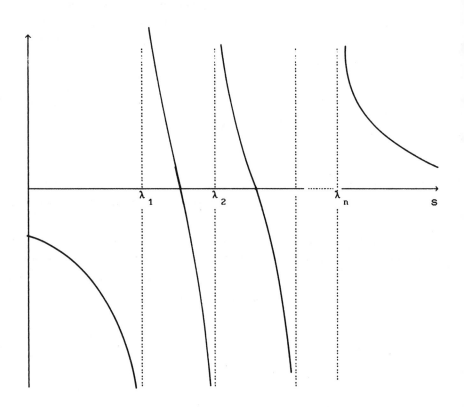

FIGURE 7:SUMMATION OF TERMS OF AUXILIARY FUNCTION

Turning now to the remaining term of equation (3.15), it can be seen that this represents a hyperbolic curve, with asymptote of the vertical axis, as shown in figure 8. of intersection of the curves shown in figures 7 and 8, as shown in figure 9.

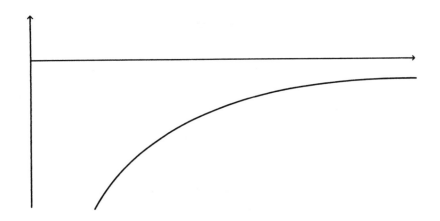

FIGURE 8: INTERSECTING FUNCTION

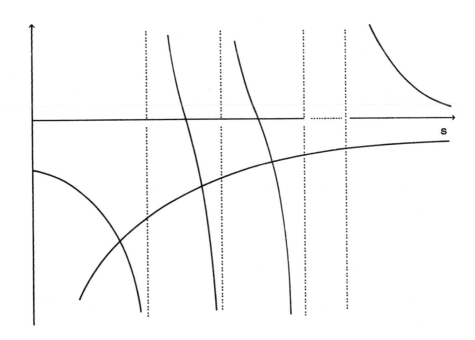

FIGURE 9: SOLUTION OF MODIFIED CHARACTERISTIC EQUATION

Weissenburger defines three entities, the intersecting function h(s) (in the current notation)

$$h(s) = \frac{-1}{\delta m \ s^2}$$

the auxiliary function g(s)

$$g(s) = \sum m_i \frac{r_i}{\left[s^2 - \lambda_i^2 \right]}$$

and their difference f(s), the modification function

$$f(s) = g(s) - h(s)$$

The solution of the eigenvalue characteristic equation therefore corresponds to the evaluation of the zeroes of the modification function. As can be seen in figure 9, the eigenvalues of the unmodified system provide bounds for those of the modified system (often described as "bracketing", see section 2.11).

Thus the solution of the characteristic equation in this form is amenable to well established equation solving algorithms, for example the Newton-Raphson method, using the eigenvalues of the initial system as starting values for the iteration. Since these algorithms are standard they will not be discussed further here. Weissenburger (1966) discusses the procedure of equation solving in some depth.

The development for stiffness modifications is similar to that described above. In this case, however, the intersecting function now is constant. It should be noted that the treatment presented here (deliberately) uses a different normalisation for the modal mass and stiffness matrices to that described by Weissenburger. Consequently, whereas Weissenburger's formulation results in a constant intersecting function for mass modifications, and a hyperbolic intersecting function for stiffness modifications, the properties of the intersecting function used here are the reverse.

Weissenburger's work (1966, 1968) contains exhaustive analysis of various special cases. This includes the treatment of systems containing equal roots. (It transpires that the major difficulty is the initial determination of the equal roots- a problem that is well known- rather than the modification procedure.) It is shown that the incorporation of combinations of mass and stiffness is possible (eg the important case of the dynamic absorber) by introducing an additional term to the auxiliary function. He proposes the incorporation of more complex modifications (ie with rank greater than unity) by the repeated application of the single rank algorithm, although he briefly discusses an alternative single step strategy. For the purposes of the current work, however, these special cases provide little additional insight into the operation and applicability of the methods, and the reader is referred to Weissenburger's original work. Of particular interest are the comments on practical numerical techniques for the equation solution phase of the method.

As has already been suggested, Weissenburger was aware of the possibility to extend the work to more complex problems, in particular to non-conservative systems and more general problems characterized by non-symmetric systems of equations.

3.2.2 Effects of Linked Stiffness

Whilst the introduction of concentrated (or "lumped") mass modifications, only affecting a single coordinate, describes a common design problem, restriction to stiffness modifications affecting only a single coordinate ("earthed" stiffeners) would be a severe limitation to the method. In practice the designer would be more likely to wish to assess the effect of linking two positions on the structure, or in strengthening an existing link.

Such modifications comprise a modification matrix containing only four non-zero terms, in the rows and columns corresponding to the proposed design change. For a unit change in stiffness between two points, the corresponding terms in the leading diagonal are 1 and the off-diagonal terms are -1.

The essential feature of such a modification term is, although the modification matrix is of order 2, the rank of the matrix is still unity. Thus it may still be factorized into the product of two vectors and only a slight variation of equation (3.11) is made.

Consider, for example, a link of stiffness δk between coordinates r and s. The form of the modification matrix is given by

$$\Delta K = \delta k\ e_{rs}\ e_{rs}^t$$

where e_{rs} has its r-th term +1, its s-th term -1 and zeroes elsewhere. In modal coordinates the modification matrix is given by $\delta k\ r_\ell\ r_\ell^t$, where $r_\ell = \Phi^t\ e_{rs}$, leading to a modified eigenvalue problem

$$\left(\sigma^2 m + \delta k\ r_\ell\ r_\ell^t + k \right)\varphi = 0$$

The solution procedure now follows the same course as described for the lumped mass or earthed stiffness modification described in section 3.2.1.

3.3 EXTENSION TO THE NON-CONSERVATIVE CASE

The extension of Weissenburger's method to dissipative systems was due to Pomazal (1969) and Pomazal and Snyder (1970). Use is made of the Duncan formulation of the eigenvalue problem, originated by Frazer, Duncan and Collar (1938) and extended by Foss (1956), Caughey (1960) and Caughey and O'Kelly (1965).

It has been noted that Weissenburger described his method initially in terms of the behaviour of continuous systems, introducing discrete modifications using the Dirac delta function. The first part of Pomazal's work re-casts Weissenburger's method for discrete systems, for example those which would result from a finite element or finite difference formulation.

Pomazal's second contribution involves the extension of the conservative case to the incorporation of damping into initially conservative systems and damping modifications into classically damped systems (ie systems where the mass, stiffness and damping distributions are diagonalized by the same modal matrix transformation). In classically (or more commonly proportionally) damped systems the eigenvalues are generally complex, in conjugate pairs, but the eigenvectors are real (and coincide with the eigenvectors of the equivalent conservative system). This is then extended to systems with more general damping distributions. The final portion of Pomazal's work provides the extension to unsymmetrical matrices and systems with repeated eigenvalues.

The current work will follow Pomazal in analysing the case for classical damping and general damping distributions in their own right rather than treating them as simply special cases of the general modification problem.

3.3.1 Modifications to Classically Damped Systems

As was described in section 2.7, classical damping implies that the modal matrix for the undamped system also diagonalizes the damping matrix. Consequently the modal matrix, and hence the eigenvectors, are identical to the undamped system (and consequently real), whereas the eigenvalues/natural frequencies become complex.

The response problem can be transformed from equation (2.3)

$$M \ddot{x} + C \dot{x} + K x = \ell(t)$$

by using the transformation to modal coordinates (equation (3.5))

$$x = \Phi q$$

to give the problem in uncoupled form

$$m \ddot{q} + c \dot{q} + k q = \Phi^t \ell(t) \qquad (3.17)$$

where c, like m and k, is a diagonal matrix.

As before, the construction of the solution to the response problem utilizes the homogeneous form of equation (3.17), the free vibration behaviour of the system, in terms of an eigenvalue problem.

$$(\lambda^2 M + \lambda C + K) \phi = 0 \qquad (3.18)$$

Now the λ_i are the solutions to the quadratic equations

$$\lambda_i^2 m_i + \lambda_i c_i + k_i = 0 \qquad (3.19)$$

which is a familiar problem from elementary single degree of freedom theory.

The solution to the response problem may be expressed (as before) in terms of the zeroes of the uncoupled equations of the individual modes, in Laplace form

$$\text{diag} \left\{ s^2 + 2 s \zeta_i \omega_i + \omega_i^2 \right\} m \, q \, (s) = \Phi \, \ell \, (s) \qquad (3.20)$$

where

$$\zeta_i = \frac{c_i}{2 \sqrt{m_i k_i}} \qquad \omega_i = \sqrt{\frac{k_i}{m_i}}$$

(again analogous results to the corresponding uncoupled single degree of freedom systems).

It can readily be seen that, as in the conservative case analysed by Weissenburger, any variation in distribution of mass, damping or stiffness will lead to a loss of orthogonality of the existing eigensystem with respect to the mass, stiffness and damping matrices. Furthermore there will be no guarantee that the eigensystem of the modified system will have proportional damping properties-indeed it is very unlikely. Nevertheless it proves possible to implement an analogous procedure to Weissenburger's method on the classically damped system.

Considering once again the effect of a mass addition at the r-th coordinate, ie the (r,r) position, leads to the modified eigenvalue problem

$$\left[\text{diag} \left\{ s^2 + 2 s \zeta_i \omega_i + \omega_i^2 \right\} m + s^2 \, \delta m \, r \, r^t \right] \varphi = 0$$
$$(3.21)$$

where r has the same significance as in equation (3.9), ie $r = \Phi^t e_r$, the rth column of Φ^t, the modal matrix of the unmodified system.

The derivation of the modification function for equation (3.21) is identical to that for the mass modification of conservative systems described in section 3.2. Thus only the result will be presented here

$$\sum_{i=1}^{n} \frac{r_i}{m_i \left(s^2 + 2 s \zeta_i \omega_i + \omega_i^2 \right)} + \frac{1}{\delta m \, s^2} = 0 \qquad (3.22)$$

The eigenvector is formed in an analogous way to that of equation (3.16), with the terms $\dfrac{r_i}{m_i \left(s^2 - \lambda_i^2 \right)}$

replaced by the terms $\dfrac{r_i}{m_i \left(s^2 + 2 s \zeta_i \omega_i + \omega_i^2 \right)}$

ie

$$\varphi = \left(\frac{r_1}{m_1 \left(s^2 + 2 s \zeta_1 \omega_1 + \omega_1^2 \right)} \right.$$

$$\left. \frac{r_2}{m_2 \left(s^2 + 2 s \zeta_2 \omega_2 + \omega_2^2 \right)} \quad \ldots\ldots \right)^t$$

It should be borne in mind that this eigenvector is defined in terms of the modal coordinate system $\{ q \}$ and that the eigenvectors of the system, in the original spatial coordinates, require the coordinate transformation from q to $æ$, given by equation (3.5): $æ = \Phi \, q$.

Whilst the derivation of the characteristic equation of the proportionally damped system, equation (3.22), is identical to that for conservative systems, its solution is not. The roots of the equation are, in general, complex, with negative real parts, and the algorithm used for its solution must meet this requirement (although there are standard solvers generally available for this common problem).

As before, the incorporation of "earthed" damping or stiffness modifications is straightforward, involving only a minor variation to equation (3.21). For a damping modification the denominator of the intersecting function becomes δc s, in place of δm s^2; For a stiffness modification the denominator is simply δk, with no frequency dependence in the intersecting function. In an analogous way to section 3.2.2 the analysis for simple linked stiffeners or dampers is straightforward, whilst the unity rank condition is maintained.

Although the initial system is proportionally damped, with real modes (eigenvectors), there is no guarantee that the modified system will retain these properties; indeed it is extremely unlikely. The characteristic equation of the modified system, equation (3.16), has complex roots and these are then substituted back to form the eigenvector of the modified system. Thus the modes of the modified system will, in general, contain complex coefficients, thereby excluding the generic properties of classically damped systems. Subsequent modifications will, therefore, require the techniques described in the next section.

3.3.2 Modifications to Symmetric Systems with General Damping Distributions: Stiffness and Damping Modifications

The construction of the key equation for Weissenburger's method, equation (3.12), exploits the orthogonality condition for conservative systems. Similarly, the construction of the characteristic equation for proportionally damped systems, equation (3.22), also utilizes the property that the modal matrix for the corresponding conservative system is capable of diagonalizing the damping matrix, ie an orthogonality condition. It follows that the analysis of more general systems by this method must also depend on the existence of a suitable orthogonality condition.

The condition required cannot be defined within the framework of the problem as defined in equation (3.2),

$$(\lambda^2 M + \lambda C + K) \phi = 0 \qquad (3.2)$$

but requires the transformation of the coordinate system to the Duncan form discussed in section 2.4. Recalling equation (2.12)

$$\begin{pmatrix} 0 & M \\ M & C \end{pmatrix} \dot{y} + \begin{pmatrix} -M & 0 \\ 0 & K \end{pmatrix} y = \begin{pmatrix} 0 \\ \ell(t) \end{pmatrix} \qquad (2.12)$$

As shown in section 2.9, the receptance representation for this system can be constructed using the orthogonality properties of the eigenvalue problem corresponding to the homogeneous form of equation (2.12). The modification analysis can be achieved by the same strategy. Writing, as in section 2.9

$$A = \begin{pmatrix} 0 & M \\ M & C \end{pmatrix} \qquad B = \begin{pmatrix} -M & 0 \\ 0 & K \end{pmatrix}$$

The eigenvalue problem $\begin{pmatrix} \lambda A + B \end{pmatrix} \psi = 0$ has eigenvectors ψ_i related to the eigenvectors of the quadratic eigenvalue problem, equation (3.12), by the formula

$$\psi_i = \begin{pmatrix} \lambda_i \phi_i \\ \phi_i \end{pmatrix} \qquad (3.23)$$

Defining the modal matrix $\Psi = \begin{pmatrix} \psi_1 & \psi_2 & \psi_3 & \cdots & \psi_{2n} \end{pmatrix}$ and quadratic products

$$a_i = \psi_i{}^t \ A \ \psi_i \qquad\qquad b_i = \psi_i{}^t \ B \ \psi_i \tag{3.24}$$

where

$$a_i = 2 \lambda_i \ \phi_i^t \ M \ \phi_i + \phi_i^t \ C \ \phi_i \tag{3.25}$$

$$b_i = -\lambda_i^2 \ \phi_i^t \ M \ \phi_i + \phi_i^t \ K \ \phi_i \tag{3.26}$$

leads to a diagonal transformation of the system matrices

$$a = \text{diag} \left(a_i \right) = \Psi^t \ A \ \Psi \tag{3.27}$$

$$b = \text{diag} \left(b_i \right) = \Psi^t \ B \ \Psi \tag{3.28}$$

This orthogonalisation procedure forms the basis of the modification method for systems with general distributions of viscous damping, first described by Pomazal (1969) and Pomazal and Snyder (1970).

Consider first a stiffness modification, of magnitude δk, as before as an "earthed" stiffener at the (r,r) position. Thus the stiffness modification matrix is given by $\Delta K = \delta k \ e_r e_r^t$, where, as before, e_r is a vector with 1 in the r-th position and zero elsewhere.

This corresponds to a modification ΔB, in B, given by

$$\Delta B = \delta k \ e_{n+r} \ e_{n+r}^t \tag{3.29}$$

where e is a vector, of order $2n \times 1$, having 1 in the $n+r$ position and zeroes elsewhere. The establishment of the characteristic equation, and the development of the eigenvectors, proceeds in an identical fashion to that which has been described in section 3.2.1. Analysis of

linked systems follows the procedure described in section
3.2.2.

The corresponding modification in b uses the
transformation to modal coordinates, but this time using
the eigenvectors of the augmented matrix, rather than the
modes of the quadratic eigenvalue problem, ie

$$\Delta b = \delta k \, \mathit{1} \, \mathit{1}^t$$

where $\mathit{1} = \Psi^t \, e_{n+r}$, as before, the (n+r)-th column of Ψ^t

The new eigenvalue problem is given by

$$\Big(s\, a + b + \Delta b \Big) \, \Upsilon = 0 \tag{3.30}$$

Since a and b are diagonal, the analysis now follows
section 3.2.1 to give a characteristic equation

$$\frac{1}{\delta k} + \sum_{i=1}^{2n} \frac{\mathit{1}_i}{a_i \left(s - \lambda_i \right)} = 0 \tag{3.31}$$

The eigenvector is given, as before, by

$$\Upsilon = \beta \left(\frac{\mathit{1}_1}{a_1 \left(s - \lambda_1 \right)} \quad \frac{\mathit{1}_2}{a_2 \left(s - \lambda_2 \right)} \quad \dots \right)^t \tag{3.32}$$

For an "earthed" damping modification $\Delta C = \delta c \, e_r \, e_r^t$ the
corresponding modification in the Duncan form is ΔA, where

$$\Delta A = \delta c \, e_r \, e_r^t$$

where e_r is a vector of order 2n×1, with 1 in the r-th
position and zeroes elsewhere. As with the stiffness
modification, the analysis now follows section 3.2.1 for

an "earthed" damper and 3.2.2 for a linked damper.

The analysis for mass modifications proved an insurmountable difficulty for Pomazal, who restricted his attention to the particular form of Duncan's method described in the literature (equation (2.12)). It can be seen, from equation (2.12), that a mass modification affects both of the matrices, A and B, a total of three separate terms. Pomazal (1969) describes an elaborate procedure for incorporation of mass modifications, registering his own dissatisfaction in the following terms

> "Because the procedure used for analysing a single modification of the mass matrix requires the investigation of four successive modifications of the augmented system matrices, the method described... lacks the formal simplicity which the stiffness and damping procedures possessed. As a result, it is possible that this mass modification procedure may not offer any advantage over a direct solution of the eigenvalue problem.... it is included here simply to complete the formal theory..."

The treatment presented here has been restricted to general structures with viscous damping, allowing the analysis to concentrate on the use of symmetric matrices. Implicitly excluded, therefore, are those systems which incorporate gyroscopic forces and constraint damping (see Meirovitch (1980)). Pomazal (1969) extended the theory to the point where it may encompass such systems, including, for example, the type of pathological systems having multiple eigenvalues with an incomplete set of eigenvectors. These will be described briefly in section 3.3.4.

3.3.3 Modifications to Unsymmetric Systems with General Damping Distributions: Hallquist's Method for Mass Modifications

Hallquist (1976), extending the work of Pomazal, established an efficient method of incorporating mass modifications, but at the expense of sacrificing the symmetry of the system equations.

Consider first the form of the augmented system used by Pomazal

$$\begin{pmatrix} 0 & M \\ M & C \end{pmatrix} \dot{y} + \begin{pmatrix} -M & 0 \\ 0 & K \end{pmatrix} y = \begin{pmatrix} 0 \\ \ell(t) \end{pmatrix} \qquad (2.12)$$

writing matrices A and B, as before

$$A = \begin{pmatrix} 0 & M \\ M & C \end{pmatrix} \qquad B = \begin{pmatrix} -M & 0 \\ 0 & K \end{pmatrix}$$

and constructing the generalized eigenvalue problem corresponding to equation (2.12)

$$\left(\lambda A + B \right) \psi = 0$$

Recall that these, eigenvectors ψ_i, are related to the eigenvectors of the quadratic eigenvalue problem, equation (3.12), by the formula

$$\psi_i = \begin{pmatrix} \lambda_i \phi_i \\ \phi_i \end{pmatrix} \qquad (3.23)$$

with the modal matrix $\Psi = \begin{pmatrix} \psi_1 & \psi_2 & \psi_3 & \cdots & \psi_{2n} \end{pmatrix}$

The inverse of A may be constructed, in partitioned form

$$A^{-1} = \begin{pmatrix} M^{-1}C^{-1}M^{-1} & M^{-1} \\ M^{-1} & 0 \end{pmatrix} \qquad (3.33)$$

Premultiplying the eigenvalue problem $\left(\lambda A + B \right) \psi = 0$

by A^{-1} gives $\left(\lambda I + A^{-1}B \right) \psi = 0$, a simple eigenvalue problem having the same eigenvalues and eigenvectors as the original eigenvalue problem.

The new eigenvalue problem is given by

$$\left\{ \lambda \begin{pmatrix} I & 0 \\ 0 & I \end{pmatrix} + \begin{pmatrix} M^{-1}C & M^{-1}K \\ -I & 0 \end{pmatrix} \right\} \psi = 0$$
$$(3.34)$$

Writing $D = \begin{pmatrix} M^{-1}C & M^{-1}K \\ -I & 0 \end{pmatrix}$ leads to the eigenvalue

problem

$$\left(\lambda I + D \right) \psi = 0. \qquad (3.35)$$

It should be noted that this eigenvalue problem retains the same right eigenvectors, ψ_i, as the original symmetric eigenvalue problem, but that the eigenvalues of the transposed eigenvalue problem, $\left(\lambda I + D^t \right) \theta$ are given by the columns of Ψ^{-t}.

In his description of the method, Hallquist establishes this form of the simple augmented eigenvalue problem using the alternative form of the Duncan transformation given by equation (2.13)

$$\begin{pmatrix} M & 0 \\ 0 & -K \end{pmatrix} \ddot{y} + \begin{pmatrix} C & K \\ K & 0 \end{pmatrix} \dot{y} = \begin{pmatrix} \ell(t) \\ 0 \end{pmatrix} \qquad (2.13)$$

In this case an identical eigenvalue problem can be derived by pre-multiplying by the matrix

$$\begin{pmatrix} M^{-1} & 0 \\ 0 & -K^{-1} \end{pmatrix}$$

As has already been mentioned, Pomazal had already developed methods for solving such systems. In this section the case of equal roots, and hence the possibility of a defective eigensystem, will be excluded. This will be discussed, however, in section 3.3.4.

The analysis for unsymmetric simple eigenvalue problems, such as equation (3.34), requires the admission of an unsymmetric modification matrix.

Denoting the augmented matrices from equation (2.13) as P and Q

$$P = \begin{pmatrix} M & 0 \\ 0 & -K \end{pmatrix} \qquad Q = \begin{pmatrix} C & K \\ K & 0 \end{pmatrix}$$

a point change δm, at the (r,r) position, will give rise to a change in the mass matrix, ΔM, with a corresponding change $\Delta P = \delta m\, e_r e_r^t$ in P.

The modified eigenvalue problem, in the original coordinates, is given by

$$\left\{ s \left[P + \delta m\, e_r e_r^t \right] + Q \right\} \vartheta = 0 \qquad (3.36)$$

The transformation to the modified unsymmetric simple eigenvalue problem is given by

$$\left\{ s \left[I + \delta m\, a_r e_r^t \right] + D \right\} \vartheta = 0 \qquad (3.37)$$

where $a_r = P^{-1} e_r$, the r-th column of P^{-1}. It can be seen that the modification matrix for the unsymmetric problem is, likewise, unsymmetric. The unsymmetric problem may now be transformed to a diagonal problem by the same method as before, diagonalising the matrices using the modal matrices, although now two different modal matrices must be used, comprising the right eigenvectors, Ψ, and transposes of the left eigenvectors, Ψ^{-1}

$$\Psi^{-1} D \Psi = \text{diag} \{ \lambda_i \} = d \qquad (3.38)$$

The equation of the modified system is now

$$\left\{ s \left(I + \delta m\, u\, v^t \right) + d \right\} \xi = 0 \qquad (3.39)$$

where $u = \Psi^{-1} a_r$, $v = e_r \Psi$ and $\vartheta = \Psi^{-1} \xi$.

$$\left\{ s\, I + d \right\} \xi = -\delta m\, s\, u\, v^t\, \xi \qquad (3.40)$$

As before, taking this equation term by term gives

$$\left(s + \lambda_i \right) \xi_i = -\delta m\, s\, u_i \sum_{k=1}^{n} v_k\, \xi_k \qquad (3.41)$$

Following the analysis of section 3.2.1 gives a characteristic equation

$$\frac{1}{\delta m\, s} + \sum_{i=1}^{2n} \frac{u_i}{\left(s - \lambda_i \right)} = 0 \qquad (3.42)$$

The eigenvector is given by

$$\xi = \left(\frac{u_1}{\left(s - \lambda_1 \right)} \quad \frac{u_2}{\left(s - \lambda_2 \right)} \quad \ldots \ldots \right)^t \qquad (3.43)$$

3.3.4 Treatment of Systems with Degenerate Eigensystems: Use of the Jordan Canonical Form

It is by no means uncommon, in works on structural dynamics, for the possibility that the problem has an incomplete eigensystem to be disregarded. This may be done either explicitly, for example as in section 2.5 of the current work ("..taking two simple..eigenvalues.."), or implicitly, when the inability to demonstrate orthogonality for repeated eigenvalues is simply not mentioned. Such an eventuality can only occur in the case when the system has repeated eigenvalues.

The liability of systems to possess multiple eigenvalues is underestimated by many analysts. Taking for example an elliptical plate, the system has two independent, ie orthogonal, first bending modes about the major and minor axes, with differing natural frequencies. If the major axis is reduced until it is equal to the minor axis then the system will still possess two natural frequencies, but these will become equal. Further, the nodal lines will become indeterminate in their direction. As a general rule of thumb, the more symmetries a structure possesses the more likely it is to have repeated natural frequencies.

For conservative systems it can be demonstrated that, even for systems with repeated eigenvalues, a full system of eigenvectors can always be found (see Bishop, Gladwell and Michaelson (1965)). This analysis depends on the positive definiteness of the system matrices and, consequently, does not extend to indefinite systems. Indeed, Pomazal uses an example of a two degree of freedom viscously damped system which possesses only one independent eigenvector.

A significant difficulty implicit with the occurrence of multiple eigenvalues is that there is no simple test to determine whether the eigensystem is defective.

The consequence of a system possessing a defective eigensystem, ie at least one eigenvalue without a corresponding eigenvector, is that it is no longer possible to find a coordinate transformation of the form of equation (3.38)

$$\Psi^{-1} D \Psi = \text{diag} \{ \lambda_i \} = d \tag{3.38}$$

Indeed the matrix Ψ no longer exists in the manner described previously, since there are insufficient proper eigenvectors (ie those corresponding to distinct eigenvalues or sets of linearly independent eigenvectors associated with multiple eigenvalues) to make up a square matrix.

However, it is possible to construct a matrix, the Jordan Canonical Form, $\tilde{\Psi}$, containing a combination of the proper eigenvectors completed with what are called generalised eigenvectors such that

$$\tilde{\Psi}^{-1} D \tilde{\Psi} = J \tag{3.44}$$

J is block diagonal

$$J = \begin{pmatrix} j_1 & & & \\ & j_2 & & \\ & & \ddots & \\ & & & j_m \end{pmatrix} \tag{3.45}$$

with Jordan blocks j_k of the form

$$j_k = \begin{pmatrix} \lambda_k & 1 & & \\ & \lambda_k & 1 & \\ & & \ddots & 1 \\ & & & \lambda_k \end{pmatrix} \tag{3.46}$$

with the λ_i in the diagonal and 1 in the superdiagonal. Superdiagonal terms only occur if the eigensystem is defective.

Using the same coordinate transformation as in the previous section the modified eigenvalue problem is given by

$$\left\{ s \left[I + \delta m \; u \; v^t \right] + J \right\} \xi = 0 \qquad (3.47)$$

Rearranging, as before

$$\left\{ s \; I + J \right\} \xi = -\delta m \; s \; u \; v^t \; \xi \qquad (3.48)$$

For eigenvalues corresponding to proper eigenvectors the analysis follows section 3.2.1, leading to individual expressions comparable with equation (3.13). For eigenvalues corresponding to generalised eigenvalues the Jordan canonical form couples at most two orthogonal vectors at each step. Consider a non-trivial Jordan block corresponding to the eigenvalue λ_k, spanning the vectors ψ_p to ψ_q. By the construction of the Jordan block ψ_q is the only proper eigenvector of the set, with $\psi_p \cdots \psi_{q-1}$ being generalised eigenvectors. (As with Gaussian elimination, the terms of the Jordan form are treated in descending order). The expression for the one proper eigenvector, ψ_q is given by

$$\left\{ s - \lambda_k \right\} \xi_q = -\delta m \; s \; u_q \sum_{i=1}^{2n} v_i \; \xi_i \qquad (3.49)$$

However, for the (q-1)-th term in the system the appropriate expression is given by

$$\left\{ s - \lambda_k \right\} \xi_{q-1} + \xi_q = -\delta m \; s \; u_{q-1} \sum_{i=1}^{2n} v_i \; \xi_i \qquad (3.50)$$

The simplification, devised by Pomazal (1969), entails the substitution of ξ_q from equation (3.50) to give

$$\left\{ s - \lambda_k \right\} \xi_{q-1} = -\delta m \ s \ \left\{ u_{q-1} + \frac{u_q}{s - \lambda_k} \right\} \sum_{i=1}^{2n} v_i \ \xi_i$$

(3.51)

The diagonalisation is now possible by defining new coefficients u_i^*, for $r \leq i \leq q-1$ where

$$u_i^* = \left\{ u_{i-1} + \frac{u_i}{s - \lambda_k} \right\}$$

(3.52)

These coefficients are now used directly in the characteristic equation, (2.42), and the eigenvector equation, (2.43).

3.3.5 Use of the Alternative Duncan Transformation to Retain Symmetric Form for Mass Modifications

It was mentioned by Hallquist (1976) that it is possible to achieve efficient reanalysis for mass modification, whilst retaining the symmetry of the system matrices. Independently, Brandon (1984c) recognized that this could be achieved more simply, by using the alternative form of the Duncan transformation defined in equation (2.13)

$$\begin{pmatrix} M & 0 \\ 0 & -K \end{pmatrix} \dot{\psi} + \begin{pmatrix} C & K \\ K & 0 \end{pmatrix} \psi = \begin{pmatrix} \ell(t) \\ 0 \end{pmatrix}$$

(2.13)

Hallquist suggests that use of the common form of the Duncan form (equation (2.12)) does not require the inversion of the stiffness matrix, and is therefore suitable for insufficiently constrained systems. However in his analysis the inverse A^{-1} is formed. As has been demonstrated above, this includes, implicitly, the inversion of the stiffness matrix. No such difficulty is encountered with the alternative Duncan form. It can be seen that, in this equivalent form to equation (2.12), the mass matrix appears only once. Because of the equivalence, between the two representations of the same system, the eigenvalues and eigenvectors are identical whichever version of the system matrices (equations (2.12) or (2.13)) is used.

3.4 USE OF EXPERIMENTAL DATA

The declared objective of the current volume is the description and classification of methods which can be used, to predict the effects of modifications, when the only data available about the structure are derived from vibration testing.

In the current chapter no mention has yet been made of the applicability of these methods to systems where only a limited amount of data are available, as would be typical in the context of experimental modal testing. This section examines the extension of these methods to experimental reanalysis. The analysis here will only consider the simple conservative problem, discussed in section 3.2.1, but readily generalizes to the viscous damping systems.

Recalling the relationship between the displacement vectors in spatial and modal coordinates, equation (3.5),

$$x = \Phi \, q \qquad\qquad (3.5)$$

there is a similar relationship between the modal matrices of the original and modified systems

$$\Phi^* = \Phi \, \Omega$$

where $\Omega = \left(\varphi_1 \ \varphi_2 \ \varphi_3 \ \cdots \ \varphi_n \right)$, the modal matrix, comprising the eigenvectors of the modified problem in modal coordinates, defined in equation (3.16), ie

$$\varphi_i = \beta \left(\frac{r_1}{m_1 \left(s_i^2 - \lambda_1^2 \right)} \quad \frac{r_2}{m_2 \left(s_i^2 - \lambda_2^2 \right)} \cdots \frac{r_n}{m_n \left(s_i^2 - \lambda_n^2 \right)} \right)^t$$

(3.53)

where s_i is the i-th root of the characteristic equation (3.15).

Examining a single column of Φ^* (an eigenvector of the modified system in spatial coordinates) gives

$$\Phi_i^* = \Phi \, \varphi_i$$

(3.54)

This product is perhaps more easily appreciated when written in series form

$$\Phi_i^* = \sum_{j=1}^{n} \Phi_j \frac{r_1}{m_1 \left(s_i^2 - \lambda_j^2 \right)}$$

(3.55)

Thus each eigenvector of the modified system is written as a linear combination of the eigenvectors of the original system weighted in terms of the coefficients of the eigenvector of the modal modification matrix.

On examination of equation (3.55) the contribution of each eigenvector of the initial system to the mode of interest of the modified system may be appreciated. The key feature is the term $(s_i^2 - \lambda_j^2)$ in the denominator of the summation. If the difference beween s_i and λ_j is large then this denominator will also be large and the contribution of the j-th mode of the initial system to the i-th mode of the modified system will be correspondingly small. This term has an analogous effect to the denominator in the spectral decomposition of the receptance matrix, described in section 2.13. As with the receptance matrix it is permissible to model the remote modes using residual coefficients.

CHAPTER 4
Modal Sensitivity Analysis

4.1 MODAL SENSITIVITY ANALYSIS: APPLICABILITY AND LIMITATIONS

It is becoming widely accepted that sensitivity analysis can be a valuable tool in structural reanalysis where (enough of) the modal properties are known, either through theoretical or experimental analysis.

Modal sensitivities are the derivatives of the modal properties of a dynamic system with respect to chosen structural variables. In the modal analysis literature there have been two primary applications. In the first case the sensitivity data are used solely as a qualitative indicator of the location and approximate scale of design changes to achieve a desired change in structural properties. The consequences of candidate design changes would then be evaluated using exact methods, either those described in the previous chapter or re-solution of the system equations.

The second strategy uses the design sensitivities directly to predict the effects of proposed structural changes. The use of sensitivities in this fashion relies on the Matrix Taylor Series expansion, with the usual implications of convergence and truncation errors. Use only of first order design sensitivities assumes implicitly that the second (and higher) order derivatives are negligible.

The use of these second order sensitivities as suitable criteria for the acceptability of first order sensitivities for predictive analysis will be discussed in some detail in the text.

In contrast to the methods described in the previous chapter, sensitivity analysis may be applied to candidate design modifications distributed across a number of degrees of freedom of the structure but, as has already been mentioned, is limited in scale.

4.1.1 Summary of Previous Work

To and Ewins (1989) have suggested that sensitivity analysis of the eigenvalue problem dates from work as early as 1846 by Jacobi. Flax (1985) notes that application in vibration analysis was due to Rayleigh in 1896 (although the 1945 edition of Rayleigh's "Theory of Sound" is most often cited in the literature).

Within the modal analysis community the earliest reference widely quoted is the work of Fox and Kapoor (1968). By using a straightforward differentiation approach they derived the first order sensitivities of both eigenvalues and eigenvectors for conservative systems. Whilst the eigenvalue sensitivity is a simple expression, the eigenvector sensitivity requires extensive matrix algebra, including the inversion of a full order matrix. It should be noted that Fox and Kapoor draw attention to the earlier publication of an equivalent eigenvalue sensitivity expression by Wittrick (1962). Fox and Kapoor examined the effects of extrapolation of eigenvalues on a simple cantilever and a simple frame structure, with rather disappointing results.

Rogers (1970) extended the work of Fox and Kapoor to unsymmetric generalised eigenvalue problems. Once again there is a suggestion that the work should be used for predictive purposes:

"... In particular, the eigenvalue derivative is seen to have profound possibilities in the optimum design of systems where dynamic response and/or dynamic stability are considerations".

Garg (1973) provided an analysis of the problem for the unsymmetric simple eigenvalue problem. The other early work commonly cited in the literature is that of Rogers (1970), Plaut and Huseyin (1976), and Nelson (1976). The advances in these works are primarily incremental, providing simplifications of the expressions devised by the earlier workers.

Latterly significant work has been presented by Vanhonacker (1980) and Whitesell (1980). Baldwin and Hutton (1985) have provided a useful survey. Perhaps the most useful single work was the Symposium held at NASA Langley Research Center in 1986 (Adelman and Hatfka 1987)), particularly the contributions by Pritchard et al (1987) and Murthy and Hatfka (1987).

4.1.2 Design Sensitivities

Modal design sensitivities are the derivatives of the eigensystem of a dynamic system with respect to those variables which are available for modification by the designer. A typical modification would be the change in diameter of a circular section. This would affect both the mass of the section, proportional to the square of the diameter, and its stiffness, which depends on the second moment of area of the section. A change in length would have a mass effect directly proportional to length, but a stiffness change depending on the cube of length. Changing material would similarly affect mass, stiffness and damping.

It would be uncommon, therefore, for practical design

problems to have design parameters that could be varied in the manner described in the last chapter.

4.2 VARIATIONAL DERIVATION OF SENSITIVITIES

In contrast to the overwhelming majority of works on sensitivity analysis, which use differentiation explicitly, a variational approach will be used here. There is a trend towards using such methods in optimization studies in structural dynamics (see for example To and Ewins (1989)). This type of approach was justified by Flax (1985) in the following terms:

"Perturbation theory and its strengths and weaknesses have been extensively explored, but the available results seem to be often overlooked,.., by investigators seeking design sensitivity factors for structural optimization analyses."

One major asset of the variational approach used here is that expansion of the modal parameters in series form explicitly forms part of the initial assumption. The derivatives required are deduced from the series rather than the converse.

For comparison equivalent expressions have been derived, using a straightforward differentiation approach, in appendix 2.

By extending the application of the perturbation method due to Lancaster (1960), the second order design sensitivities of both eigenvalues and eigenvectors, for self adjoint systems, ie encompassing passive, non-gyroscopic systems, were derived previously by the author (Brandon (1984c)). As shown by Flax (1985), the analysis was restricted to stiffness modifications (see also Brandon (1985)). In the current work the method is extended both to mass and damping modifications and also to non self-adjoint systems, admitting gyroscopic and active damping mechanisms.

4.2.1 Assumptions

The derivation presented here is slightly less general than that suggested in section 4.1.2. The analysis applies to the non-symmetric, generalised eigenvalue problem

$$\left(\lambda A + B \right) \psi = 0 \tag{4.1}$$

but assumes that the system has discrete eigenvalues. For application to more general systems (involving the use of the Jordan Canonical form) the reader is referred to Collar and Simpson (1987). Under these assumptions an orthogonal transformation, similar to equations (3.24), (3.27) and (3.28) always exists:

$$\theta_i^t A \psi_i = a_i \tag{4.2}$$

$$\theta_i^t B \psi_i = b_i \tag{4.3}$$

$$\theta_j^t A \psi_i = 0 \tag{4.4}$$

$$\left. \begin{array}{l} \\ \\ \end{array} \right\} j \neq i$$

$$\theta_j^t B \psi_i = 0 \tag{4.5}$$

where θ_i^t denotes the i-th left eigenvector.

In section 4.1.2, however, design sensitivities are defined in terms of parameters available to the designer. As described in that section, a change in mass or stiffness may not be directly proportional to the variation in the design parameter. In the following analysis it will be assumed that the changes in mass, stiffness and damping distribution are directly proportional to the design change. The method of extension to the more general case

will be indicated at the appropriate point in the text.
There are two motivations for this approach. Firstly, the
algebra becomes significantly more complicated without
providing a concomitant enhancement of insight into the
problem. Secondly, the non-proportional element only
affects the second and higher order sensitivities. It is
argued in the text that the knowledge derived about higher
order sensitivities is primarily of a qualitative nature.

4.2.2 Derivation

Under the assumptions made above, the eigenvalue problem
of the perturbed system is given by

$$\left\{ \lambda_i^* \left(A + \varepsilon \, \alpha \right) + \left(B + \varepsilon \, \beta \right) \right\} \psi_i^* = 0 \qquad (4.6)$$

Following Lancaster (1960), since the characteristic
equation of the modified problem is algebraic in ε, the
eigenvalues λ_i^* and eigenvectors ψ_i^* of the modified system
can be expressed as a series in ε:

$$\lambda_i^* = \lambda_i^{(0)} + \varepsilon \, \lambda_i^{(1)} + \varepsilon^2 \, \lambda_i^{(2)} + \ldots \qquad (4.7)$$

$$\psi_i^* = \psi_i^{(0)} + \varepsilon \, \psi_i^{(1)} + \varepsilon^2 \, \psi_i^{(2)} + \ldots \qquad (4.8)$$

Substituting these expressions into the characteristic
equation of the modified system, equation (4.6), gives:

$$\left\{ \left(A + \epsilon \ \alpha \right) \left(\lambda_i^{(0)} + \epsilon \ \lambda_i^{(1)} + \epsilon^2 \ \lambda_i^{(2)} \right) + \right.$$

$$\left. \left(B + \epsilon \ \beta \right) \right\} \left(\psi_i^{(0)} + \epsilon \ \psi_i^{(1)} + \psi_i^{(2)} \right) = 0$$

$$(4.9)$$

This expression can be expanded in powers of ϵ, and the linear independence properties exploited, to give closed form expressions for the perturbation coefficients.

In ϵ^0:

$$\left(\lambda_i^{(0)} \ A + B \right) \psi_i^{(0)} = 0 \qquad\qquad (4.10)$$

corresponding to the properties of the unperturbed system.

In ϵ^1:

$$A \ \lambda_i^{(0)} \ \psi_i^{(1)} + A \ \lambda_i^{(1)} \ \psi_i^{(0)} + \alpha \ \lambda_i^{(0)} \psi_i^{(0)} +$$

$$B \ \psi_i^{(1)} + \beta \ \psi_i^{(0)} = 0 \qquad\qquad (4.11)$$

In ϵ^2:

$$A \ \lambda_i^{(0)} \ \psi_i^{(2)} + A \ \lambda_i^{(1)} \ \psi_i^{(1)} + A \ \lambda_i^{(2)} \ \psi_i^{(0)}$$

$$+ \alpha \ \lambda_i^{(0)} \ \psi_i^{(1)} + \alpha \ \lambda_i^{(1)} \ \psi_i^{(0)} + B \ \psi_i^{(2)} + \beta \ \psi_i^{(1)} = 0$$

$$(4.12)$$

Again following Lancaster (1960), an orthogonality condition may be imposed between the left eigenvectors and the perturbation coefficients

$$\theta_i^t \ A \ \psi_i^{(1)} = \theta_i^t \ A \ \psi_i^{(2)} = \theta_i^t \ B \ \psi_i^{(1)} = \theta_i^t \ B \ \psi_i^{(2)} = 0$$

(4.13)

Premultiplying equation (4.11) by θ_i^t and applying these orthogonality conditions gives

$$\lambda_i^{(1)} \ \theta_i^t \ A \ \psi_i^{(0)} + \lambda_i^{(0)} \ \theta_i^t \ \alpha \ \psi_i^{(0)} + \theta_i^t \ \beta \ \psi_i^{(0)} = 0$$

(4.14)

Rearranging to give $\lambda_i^{(1)}$

$$\lambda_i^{(1)} = - \left\{ \frac{\theta_i^t \ \left(\lambda_i^{(1)}\alpha + \beta \right) \ \psi_i^{(0)}}{\theta_i^t \ A \ \psi_i^{(0)}} \right\}$$

(4.15)

Exploiting the orthogonality conditions, equation (4.13), the first variation in eigenvector $\psi_i^{(1)}$ may be expressed in terms of the unperturbed eigenvectors

$$\psi_i^{(1)} = \sum_{\substack{j=1 \\ j \neq i}}^{n} p_{ij} \ \psi_j^{(0)}$$

(4.16)

Premultiplying equation (4.11) by θ_j^t and exploiting orthogonality gives

$$\lambda_i^{(0)} \ \theta_j^t \ A \ \psi_i^{(1)} + \lambda_i^{(1)} \ \theta_j^t \ A \ \psi_i^{(0)} + \lambda_i^{(0)} \ \theta_j^t \ \alpha \ \psi_i^{(0)}$$

$$+ \ \theta_j^t \ B \ \psi_i^{(1)} + \theta_j^t \ \beta \ \psi_i^{(0)} = 0$$

(4.17)

Expanding this expression in terms of p_{ij} gives

$$p_{ij} = - \left\{ \frac{\theta_j^t \left(\lambda_i^{(0)} \alpha + \beta \right) \psi_i^{(0)}}{\theta_j^t \left(\lambda_i^{(0)} A + B \right) \psi_i^{(0)}} \right\} \qquad (4.18)$$

The second variations are derived in the same manner as that just described. Pre-multiplying equation (4.12) by θ_i^t and exploiting the orthogonality conditions, equation (4.13), gives

$$\lambda_i^{(2)} \theta_i^t A \psi_i^{(0)} + \lambda_i^{(0)} \theta_i^t \alpha \psi_i^{(1)} + \lambda_i^{(1)} \theta_i^t \alpha \psi_i^{(0)}$$

$$+ \theta_i^t \beta \psi_i^{(1)} = 0 \qquad (4.19)$$

Rearranging for $\lambda_i^{(2)}$:

$$\lambda_i^{(2)} = - \left\{ \frac{\theta_i^t \left(\lambda_i^{(0)} \alpha + \beta \right) \psi_i^{(1)} + \lambda_i^{(1)} \theta_i^t \alpha \psi_i^{(0)}}{\theta_i^t A \psi_i^{(0)}} \right\}$$

$$(4.20)$$

This expression coincides (with slightly different notation) with the expression derived by Flax (1985).

Once again the eigenvector variation may be expressed in terms of the orthogonality conditions, equation (4.13)

$$\psi_i^{(2)} = \sum_{\substack{j=1 \\ j \neq i}}^{n} q_{ij} \psi_j^{(0)} \qquad (4.21)$$

The expression for the variation in eigenvector requires equation (4.12) to be pre-multiplied by θ_j^t and the orthogonality conditions (4.13) used

$$\lambda_i^{(0)} \; \theta_j^t \; A \; \psi_i^{(2)} \;\; + \lambda_i^{(1)} \; \theta_j^t \; A \; \psi_i^{(1)} \; + \lambda_i^{(0)} \; \theta_j^t \; \alpha \; \psi_i^{(1)}$$

$$+ \lambda_i^{(1)} \; \theta_j^t \; \alpha \; \psi_i^{(0)} \; + \theta_j^t \; B \; \psi_i^{(2)} \; + \theta_j^t \; \beta \; \psi_i^{(1)} \; = \; 0$$

$$(4.22)$$

Substituting equation (4.21) into equation (4.22) allows the q_{ij} to be derived

$$q_{ij} \; = \; -\left\{ \frac{\theta_j^t \; \left[\lambda_i^{(0)}(A + \alpha) + (B + \beta)\right] \; \psi_i^{(1)} + \lambda_i^{(1)}\theta_j^t \; \alpha \; \psi_i^{(0)}}{\theta_j^t \; \left[\lambda_i^{(0)}A + B\right] \; \psi_j^{(0)}} \right\}$$

$$(4.23)$$

4.3 RELATIONSHIP BETWEEN VARIATIONAL AND DIFFERENTIAL SENSITIVITY COEFFICIENTS

It has already been mentioned that the derivation of sensitivities results in equivalent expressions whichever method is used. Flax (1985) describes this equivalence in the following terms:

"Rayleigh's methods are well known in the literature.
.. His formulation is general and valid whether the eigenvalue problem is in matrix, differential, or integral equation form".

The equivalence may be established by comparing the power series for the variational forms, equations (4.7) and (4.8), with the corresponding Taylor series, denoting a change δv in a design varible v, which, consistent with the discussion in section 4.1.2, may comprise a combination of mass, stiffness or damping.

$$\lambda_i^* = \lambda_i + \delta v \frac{\partial \lambda_i}{\partial v} + \frac{(\delta v)^2}{2} \frac{\partial^2 \lambda_i}{\partial v^2} + \qquad (4.24)$$

$$\psi_i^* = \psi_i + \delta v \frac{\partial \psi_i}{\partial v} + \frac{(\delta v)^2}{2} \frac{\partial^2 \psi_i}{\partial v^2} + \qquad (4.25)$$

This may be compared, term by term, with the variational power series, equations (4.7) and (4.8)

$$\lambda_i^* = \lambda_i^{(0)} + \epsilon \lambda_i^{(1)} + \epsilon^2 \lambda_i^{(2)} + \ldots \qquad (4.7)$$

$$\psi_i^* = \psi_i^{(0)} + \epsilon \psi_i^{(1)} + \epsilon^2 \psi_i^{(2)} + \ldots \qquad (4.8)$$

leading to the expressions

$$\delta v \frac{\partial \lambda_i}{\partial v} = \epsilon \lambda_i^{(1)} \qquad (4.26)$$

$$\frac{(\delta v)^2}{2} \frac{\partial^2 \lambda_i}{\partial v^2} = \epsilon^2 \lambda_i^{(2)} \qquad (4.27)$$

$$\delta v \frac{\partial \psi_i}{\partial v} = \epsilon \psi_i^{(1)} \qquad (4.28)$$

$$\frac{(\delta v)^2}{2} \frac{\partial^2 \psi_i}{\partial v^2} = \epsilon^2 \psi_i^{(2)} \qquad (4.29)$$

Recalling that α and β were the increments, due to a unit change in the design variable v, in A and B respectively, leads to expressions for the first order design

84

sensitivities in differential form:

$$\frac{\partial \lambda_i}{\partial v} = \frac{-\theta_i^t \left\{ \frac{\partial B}{\partial v} + \lambda_i \frac{\partial A}{\partial v} \right\} \psi_i}{\theta_i^t A \psi_i}$$

(4.30)

$$\frac{\partial \psi_i}{\partial v} = \sum_{j=1, j \neq i}^{n} p_{ij} \psi_j$$

(4.31)

where:

$$p_{ij} = \frac{-\theta_j^t \left\{ \frac{\partial B}{\partial v} + \lambda_i \frac{\partial A}{\partial v} \right\} \psi_i}{\left(\lambda_i - \lambda_j \right) \theta_j^t A \psi_j}$$

(4.32)

These expressions are consistent with those derived by Rogers (1970).

The second order design sensitivities are obtained in a similar manner. The eigenvalue sensitivity is given by the expression:

$$\frac{\partial^2 \lambda_i}{\partial v^2} = \frac{-2 \left[\frac{\partial \lambda_i}{\partial v} \theta_i^t \frac{\partial A}{\partial v} \psi_i + \theta_i^t \left\{ \frac{\partial B}{\partial v} + \lambda_i \frac{\partial A}{\partial v} \right\} \frac{\partial \psi_i}{\partial v} \right]}{\theta_i^t A \psi_i}$$

(4.33)

As previously, the eigenvector sensitivity is expressed in terms of the sum of a series comprising the orthogonal set of eigenvectors:

$$\frac{\partial^2 \psi_i}{\partial v^2} = \sum_{\substack{j=1 \\ j \neq i}} q_{ij} \, \psi_j \tag{4.34}$$

where the q_{ij} are given by:

$$q_{ij} = \frac{-2\left[\frac{\partial \lambda_i}{\partial v} \, \theta_j^t \, \frac{\partial A}{\partial v} \, \psi_i + \theta_j^t \left\{ \frac{\partial B}{\partial v} + \frac{\partial \lambda_i}{\partial v} A + \lambda_i \frac{\partial A}{\partial v} \right\} \frac{\partial \psi_i}{\partial v} \right]}{\left(\lambda_i - \lambda_j \right) \theta_j^t \, A \, \psi_j} \tag{4.35}$$

4.4 SIGNIFICANCE OF HIGHER ORDER SENSITIVITIES

In section 4.1 two primary applications of sensitivity analysis were described. Firstly, the sensitivity data are used qualitatively to indicate the location and approximate scale of design changes to achieve a desired change in structural properties. In the second case the design sensitivities are used directly (using a truncation of the matrix Taylor series) to predict the effects of proposed structural changes.

In either case there is considerable potential benefit in computing the higher order sensitivities. In the context of the qualitative use of the sensitivity coefficients the higher order sensitivities give a measure of the degree to which the solution is purely local. This will give the analyst a measure of confidence in the applicability, or otherwise, and probable quality of solution, of the exact methods described in chapters 3 and 5. For extrapolation, the quality of the computed solution can be assessed by considering the convergence and truncation errors of the Matrix Taylor Series. The intuitive appeal of the first order sensitivities, particularly the eigenvalue sensitivity, is, at first sight, compelling. The first order sensitivity can be computed solely on the basis of knowledge of the distribution of the candidate design modifications and the associated eigenvector. This eigenvector may be obtained analytically or experimentally. The first signs of doubt should start to occur when the first order eigenvector sensitivity is considered. The key consideration is the term $\left(\lambda_i - \lambda_j\right)$ in the denominator, which demonstrates that eigenvector sensitivities may become large, and potentially ill-conditioned, if errors occur in the calculation (or experimental identification) of λ_i or λ_j, for adjacent modes with close natural frequencies. The equation will degenerate for $\lambda_i = \lambda_j$, although there are methods in the literature which can accommodate this problem (unless the set of proper eigenvectors is incomplete), see for example Collar and Simpson (1987).

Turning now to the second order eigenvalue sensitivity, it can be seen that this incorporates the first order eigenvector sensitivity of the mode of interest. As has been described in the previous paragraph, this will be (potentially) poorly behaved if there is another close mode.

Continuing the process of examining higher order sensitivities, the second eigenvector sensitivity contains a factor $\left(\lambda_i - \lambda_j\right)^2$ in its denominator. Thus whatever condition (good or bad) is present in the first order eigenvector sensitivity will be even more significant in the second order eigenvector sensitivity. If the derivation process of section 4.3 is reconsidered, it can be seen that a recurrence relation can be built, where each new eigenvalue sensitivity has the same order of $\left(\lambda_i - \lambda_j\right)^n$ as the previous eigenvector sensitivity, and each new eigenvector sensitivity introduces one higher power of $\left(\lambda_i - \lambda_j\right)$. In addition, it should be noted that this term alternates in sign.

4.4.1 Physical and Numerical Constraints on the Taylor Series

The overwhelming majority of the published literature treats the sensitivities as autonomous numerical entities. This implicitly sacrifices valuable information which may be derived by consideration of the physical and numerical constraints of the problem. As has been seen in section 2.11, and exploited extensively in chapter 3, a rank one modification has a limited, and predictable, effect on the eigenvalues of the system, which are constrained by the bracketing theorems. (See also Skingle and Ewins (1989)).

Whilst the candidate design changes considered in modal sensitivity analysis are no longer restricted to single rank modifications, the new restriction to small magnitude modifications means that *de facto*, rather than *de jure*, bracketing conditions apply. Wang, Heylen and Sas (1987) used analysis similar to that presented in section 3.2.1 to compare the applicability of procedures based on Taylor series and exact methods for modification analysis. The bracketing theorems can be established by consideration of the continuity of the Rayleigh quotient. To and Ewins (1989) have devised a modification procedure which is explicitly based on the use of the Rayleigh quotient to devise sensitivities.

4.4.2 Applicability Criteria for Modal Sensitivities

Examination of equations (4) and (7) leads to a simple acceptability criterion for the use of a truncated Taylor series extrapolation. Consider for example the eigenvalue Taylor series, equation (4.24):

$$\lambda_i^* = \lambda_i + \delta v \frac{\partial \lambda_i}{\partial v} + \frac{(\delta v)^2}{2} \frac{\partial^2 \lambda_i}{\partial v^2} + \ldots$$

The methods of derivation of the sensitivities leads to the conclusion that the rth eigenvector sensitivity contributes to the (r+1)th eigenvalue sensitivity. In addition each eigenvector sensitivity introduces one higher power of the ratio

$$\rho_{ij} = \frac{- \theta_j^t \left\{ \frac{\partial B}{\partial v} + \lambda_i \frac{\partial A}{\partial v} \right\} \psi_i}{\left(\lambda_i - \lambda_j \right) \theta_j^t A \psi_j} \qquad (4.36)$$

Analogy with the Cauchy ratio test leads to the belief that the convergence of the Taylor series (4.25) is likely to depend (at least in part) on the ratio:

$$\gamma_{ij} = \left| \delta v \sum_{\substack{j=1 \\ j \neq i}}^{n} \frac{\theta_j^t \left\{ \frac{\partial B}{\partial v} + \lambda_i \frac{\partial A}{\partial v} \right\} \psi_i}{\theta_i^t \left\{ \frac{\partial B}{\partial v} + \lambda_i \frac{\partial A}{\partial v} \right\} \psi_i} \rho_{ij} \right| \qquad (4.37)$$

and that early truncation is acceptable if $\gamma_{ij} \ll 1$.

Difficulties may be expected if either $\theta_j^t \left\{ \frac{\partial B}{\partial v} + \lambda_i \frac{\partial A}{\partial v} \right\} \psi_i$

is large   or $(\lambda_i - \lambda_j)$ is small.

To and Ewins (1989) discuss a different criterion for the acceptability of extrapolation using first order sensitivities, based on the work of Stewart (1973). Whilst Stewart's work was based on the simple eigenvalue problem, To and Ewins have extended the condition estimator to the generalised eigenvalue problem. The description of the extended method by To and Ewins is so brief that it is not possible to discuss it in the current work.

4.4.3 Numerical Example

The illustration chosen is an extremely simple three degree of freedom spring mass system (Fig. 10).   The initial model has a permutation symmetry (i.e. K and M are unchanged by renumbering x_2 and x_3).   The second and third eigenvalues are close together and symmetry leads to a lack of relative movement in a pair of coordinates in each mode.

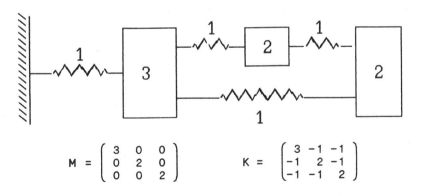

$$M = \begin{pmatrix} 3 & 0 & 0 \\ 0 & 2 & 0 \\ 0 & 0 & 2 \end{pmatrix} \qquad K = \begin{pmatrix} 3 & -1 & -1 \\ -1 & 2 & -1 \\ -1 & -1 & 2 \end{pmatrix}$$

FIGURE 10: TEST SYSTEM

For comparison of results an accurate standard eigensolver, using Householder's method and QL transformation, is used (based on procedures TRED2, TQL2 from Wilkinson and Reinsch (1971).Consider the effect of proposed mass changes at x_1 and x_2. Fig. 11 compares the actual behaviour of the eigenvalue with that predicted by first order extrapolation.

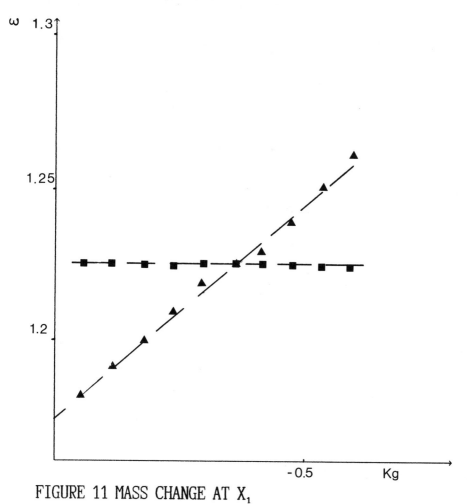

FIGURE 11 MASS CHANGE AT X_1

For changes at x_1 it can be seen that the first order approximation closely predicts actual system behaviour.

In contrast, in figure 12, changes at x_2 reveal gross discrepancies between actual and predicted system behaviour.

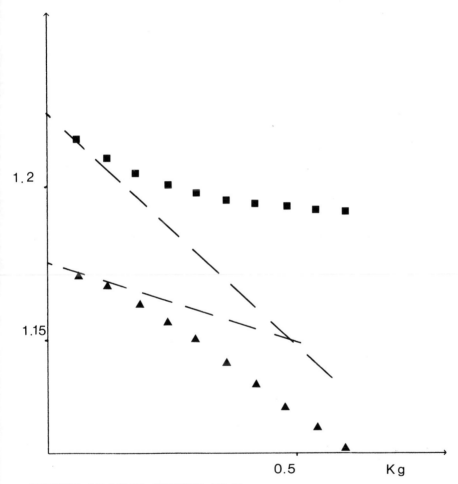

FIGURE 12 MASS CHANGE AT X_2

The explanation for this apparently paradoxical system behaviour is that x_1 is a node for the third mode, hence the ρ_{ij} are zero. At x_2 the coupling effect between the modes, represented by the quality ratio ρ_{ij}, is large and the difference between the eigenvalues is small. Computing the eigenvector sensitivities:

$$\frac{\partial \psi_1}{\partial m_2} = 0.007 \quad \psi_2 \quad + \quad 0.018 \quad \psi_3$$

$$\frac{\partial \psi_2}{\partial m_2} = -0.275 \quad \psi_1 \quad - \quad 1.523 \quad \psi_3$$

$$\frac{\partial \psi_3}{\partial m_2} = 0.514 \quad \psi_1 \quad + \quad 1.740 \quad \psi_2$$

From equation (4.33) we are led to expect, therefore, the observed significant curvature in the eigenvalue locus as m_2 changes.

For comparison, the corresponding frequency sensitivities are:

$$\frac{\partial \omega_1}{\partial m_2} = - 0.030 \quad i$$

$$\frac{\partial \omega_2}{\partial m_2} = -0.044 \quad i$$

$$\frac{\partial \omega_3}{\partial m_2} = -0.153 \quad i$$

NB $\omega_i^2 = \lambda_i$ $\qquad \frac{\partial \lambda_i}{\partial v} = 2 \omega_i \frac{\partial \omega_i}{\partial v}$

Fig. 13 shows the effect of two improvement strategies. In

the first case it is recognised that the modal mass m_i will change and hence this is updated at each step. Secondly the eigenvector sensitivities are used to update the whole eigensystem.

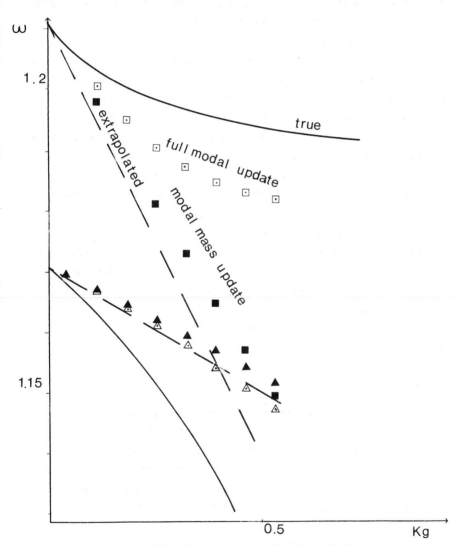

FIGURE 13: COMPARISON OF UPDATE STRATEGIES

As can be seen from Fig. 13, both methods give some improvement in the upper eigenvalue estimate with the (considerably) more expensive eigenvector method giving correspondingly better results. In the case of the lower eigenvalue there is no improvement and the modal update method actually shows a deterioration.

The nature of the Taylor series expansion guarantees that refining the step size will improve the convergence of the method (At least it will until computational discretisation errors become significant.) Consequently if a case is found where a poorly behaved solution is expected (by the test of equation (4.37)), then step size refinement is an acceptable strategy. In applications where poorly conditioned eigensystems are encountered it is possible that equation (4.37) may be implemented adaptively at each extrapolation step.

In a wide range of practical problems, however, the difficulties described here are seldom encountered. Since, however, in general, eigenvector sensitivities are available it is a relatively cheap and straightforward task to compute the acceptability parameter of equation (4.37). For well behaved sensitivities this will give an added measure of confidence in the results.

4.4 COMPUTATIONAL PROCEDURES

The sensitivity expressions, equations (4.30)-(4.35), appear to be daunting in their complexity, apparently requiring a substantial number of matrix computations. The formation of the second order sensitivities is, however, quite straightforward and utilises coefficients previously calculated (or measured) as part of the computation of the first order sensitivities.

The computational effort involved in their incorporation into the second order truncated Taylor series is, however, considerable and for many purposes will be regarded as prohibitive.

The primary value of these expressions is in the formation of condition estimators for assessing the applicability of first order sensitivity extrapolations. The condition estimator proposed here has been developed using heuristic arguments and no claim is made as to its general applicability. With the incomplete description of the method devised by To and Ewins (1989) it is not possible to provide a detailed analysis of the method, although their approach seems promising.

There is evidence in the literature for the belief that gross extrapolation using first order sensitivities is in general use. The availability of condition estimators which assess the applicability of the more simplistic models must be considered to represent a significant advance.

CHAPTER 5

Exact Methods for Response Prediction

5.1 EXACT MODIFICATION METHODS FOR RESPONSE IN CONTEXT: SUBSTRUCTURES AND SUPER-ELEMENTS

The methods which will be described in the current chapter fall into a class between two general analytical categories. Firstly, they are a special case of the techniques commonly described as substructure methods. There is a substantial body of literature related to this area. In the current context, that of the relationship between analytical and experimental models, the volume by Martinez and Miller (1985) gives a balanced overview. Of particular interest are the papers by Craig (1985) and Niedbal (1985). From the opposite point of view, the finite element community would probably classify the modification matrices used here as "super-elements". Typical of the type of modification which would fall readily into this category would be the type of beam element used by Elliott and Mitchell (1987).

The view taken in the current work is that the modifications used are sufficiently specialised that they do not fit easily into either of the other categories. In substructure synthesis the substructures which are assembled are typically of similar size and complexity. The matrix manipulations involved use large matrices of significant rank. (One of the key problems is the procedure for estimating suitable modal truncations, and hence the rank, of the substructures prior to assembly of the composite structure). The theme of the analysis presented here is based on an assumption that the modification matrices are sufficiently small, either in rank or magnitude, that the internal resonant characteristics of the modifications contribute little to the global properties of the modified structure.

Classifying the modifications as super-elements would only be of value if a suitable theoretical model were already available, for example as described by Elliott and Mitchell (1987) or for the model adjustment problems discussed in section 2.13. This cannot be guaranteed in experimental structural dynamics and is explicitly excluded in the declared scope of the current volume. The power of the finite element method derives, to a substantial extent, on simple and repetitive topology of the elements. Such a methodology would be a significant disadvantage in the methods described in this volume.

5.2 TWO APPROACHES: PSEUDO-FORCE AND KRON'S METHOD

The objective of this section is to demonstrate the equivalence of two methods previously treated as distinct in the literature. It has been argued previously by the author and co-workers (Sadeghipour et al (1985), Brandon et al (1988)) that the Kron formulation leads to a representation of the problem which retains advantageous numerical properties of the equations.

The pseudo-force method is analogous to the compatibility method in static structural analysis. At positions on the structure where a designer wishes to have control of the design variables internal constraint forces are replaced by external forces with the same dynamic properties.

Consider the receptance formulation of the unmodified

system, equation (2.22)

$$S \, \pmb{\mathscr{x}} \; = \; \left(s^2 \, M \; + \; s \, C \; + \; K \right) \pmb{\mathscr{x}} \;\; = \pmb{\mathscr{f}} \, (s)$$

The inverse of S is the receptance matrix R, ie

$$R \; = \; S^{-1} \; = \; \left(s^2 \, M \; + \; s \, C \; + \; K \right)^{-1}$$

The methods described in the current chapter seek an exact formula to evaluate the receptance matrix of the modified system in terms only of the receptance matrix of the initial system and the modification. Since the receptance matrix may be derived from experimental testing (or at least a sufficiently good approximation) the application to structures with incomplete design data is possible. It should be borne in mind that the receptance matrix is a function of frequency and is often only of interest at isolated frequencies or within a small range.

Taking a general change ΔS in the dynamic stiffness matrix S, which may be a combination of mass, stiffness or damping changes, ie

$$\Delta S \;\;\; = \; \left(s^2 \, \Delta M \; + \; s \, \Delta C \; + \; \Delta K \right) \tag{5.1}$$

leads to a new receptance matrix R^* given by

$$R^* \; = \; \left(S \; + \; \Delta S \right)^{-1} \tag{5.2}$$

The pseudo-force method has much in common with the exact modal modification analysis described in chapter 3. Once again the internal constraint forces, due to the combination of the modification and the modified displacement vector $\pmb{\mathscr{x}}^*$, are taken to the right hand side of the equation

$$\left(s^2 M + s C + K \right) \text{\ae}^* = \ell (s) - \Delta S \text{\ae}^* \qquad (5.3)$$

The method, as with the exact modal method, relies on the low rank of ΔS to solve the equation

$$\text{\ae}^* = R^* \ell \qquad (5.4)$$

Derivations and application of the method are widely published in the literature (see for example Hammill and Andrew (1973) and Mahalingam (1975a,b)) and will not be pursued further here, once the equivalence with Kron's method is established.

Kron's work was originally devised for analysing the dynamic characteristics of electrical networks (Kron (1963)). He recognised at an early stage that the techniques he devised potentially had much wider application and applied them to a diverse range of topics, including linear programming, particle physics and, most importantly in the current context, structural analysis. Despite the vigorous advocacy of Simpson (Simpson and Tabarrok (1968), Simpson (1980) and Collar and Simpson (1987)) Kron's methods have been somewhat neglected by the structural dynamics community. This is perhaps due to the rather difficult notation (developed by Kron in the absence of contemporary interest by mathematicians) but also to the rather metaphysical nature of the arguments. As was suggested in section 1.3 Noble (1969) recognised the importance of Kron's key theorem:

For a dynamic stiffness matrix S, with associated receptance matrix $R = S^{-1}$, if a localised modification ΔS of the form $\delta v \, U \, V$, where δv is a variable scalar, U is n×r and V is r×n, then the new receptance matrix,

$$R^* = \left(S + \Delta S\right)^{-1} \text{ is given by}$$

$$R^* = R - \delta v \, R \, U \, W^{-1} V \, R \qquad (5.5)$$

where

$$W = I_r + \delta v \, V \, R \, U \qquad (5.6)$$

subject to the condition $| \, W \, | \neq 0$ (ie W non-singular). This theorem is proved in Appendix 3.

Instead of the inversion of an n×n matrix $\left(S + \Delta S\right)$ it has been possible to transform the problem to an inversion (explicit or implicit) of the r×r matrix W combined with a series of matrix multiplications. Since the matrix multiplications are $O(nr^2)$ and inversions are $O(n^3)$ the method provides computational advantages for problems of only moderate scale.

In the case of a modification which only affects only a single term of S, eg a mass modification or earthed stiffness or damper, then the modification δv at the i-th coordinate (which may be frequency dependent if it includes inertia or damping) will give rise to a change $\delta v \, e_i e_i^t$ in S and the new receptance matrix is given by

$$R^* = R - \frac{\delta v \, c_i c_i^t}{1 + r_{ii} \, \delta v} \qquad (5.7)$$

where r_{ii} is the (i,i) element of R and c_i is the i-th column of R. In this case the modification to the receptance matrix comprises only the multiplication of a scalar by a row-column product.

5.3 A DIGRESSION: SERIES EXPANSION FOR THE RECEPTANCE MATRIX OF MODIFIED SYSTEMS

Chapter 4 considered a variational form of a series expansion and the matrix Taylor series. In the current context a different aspect of the polynomial series form is appropriate. This is the binomial expansion of the receptance of the modified system viewed as the inverse of the modified dynamic stiffness matrix

$$R^* = \left(S + \Delta S\right)^{-1} \tag{5.8}$$

This may be rearranged

$$R^* = \left(I + S^{-1}\Delta S\right)^{-1} S^{-1} \tag{5.9}$$

Recalling that $S^{-1} = R$, this may be expanded in terms of the binomial series

$$R^* = \left(I - (R\ \Delta S) + (R\ \Delta S)^2 - \ldots\right) R \tag{5.10}$$

subject to the convergence criterion, expressed in the least squares norm

$$\| R\ \Delta S \|_2 < 1 \tag{5.11}$$

This expression, as with other methods in the book, contains only information about the known response properties of the system (which may be approximated using experimental measurements) and the modification matrix. It should be noted that there are other matrix norms which could be used, for example the unity and infinity norms $\| \circ \|_1$ and $\| \circ \|_\infty$ which are often simpler (and cheaper) to compute than the norm chosen here (see Noble and Daniel (1977)). The particular importance of the

$\| \circ \|_2$ norm is its relationship with the eigenvalues of the matrix (specifically the largest eigenvalue).

Following the method described in section 5.2, computational advantages may be gained by exploiting the small rank of ΔS, for example using the factorisation $\Delta S = \delta v \; U \; V$.

The convergence criterion provides valuable information with respect to the applicability of the binomial expansion, for response modification analysis, which will be used again in chapter 6.

A more conservative convergence can be derived by applying the well known (see for example Noble and Daniel (1977)) Schwarz inequality to equation (5.11)

$$\| \; R \; \|_2 \; \| \; \Delta S \; \|_2 \; < \; 1 \qquad\qquad (5.12)$$

This expression implies that the binomial series will fail to converge if either $\| \; R \; \|_2$ or $\| \; \Delta S \; \|_2$ is large. Because the $\| \circ \|_2$ norm is, for square matrices, defined in terms of the largest eigenvalue, and in turn the largest eigenvalue is related to the resonant characteristics of the system, the convergence of the series will fail when $\| \; R \; \|_2$ is large, ie in the region of a resonance, or when $\| \; \Delta S \; \|_2$ is large, ie if the proposed modification is of too large a magnitude.

An interesting special case is the undamped dynamic absorber where $\| \; \Delta S \; \|_2 \longrightarrow$ minimum as $s \longrightarrow \omega_{\Delta S}$. Thus in the immediate neighbourhood of the resonant frequency of the absorber the Taylor series will converge but outside a very limited range the norm of the receptance matrix will dominate. As described by Weissenburger (1966) the addition of the absorber requires the augmentation of the original system with an additional degree of freedom which has zero displacement in the initial system.

Whilst equation (5.12) provides a sufficient, but not necessary, condition for convergence (whereas the more stringent requirement, equation (5.11), is a necessary condition) there will be few practical design problems where the distinction is of any importance. The interpretation of this more conservative convergence criterion is that the binomial series is least likely to converge when $\parallel R \parallel_2$ is large, a condition which occurs in the region of a resonance.

5.4 PRACTICAL LIMITATIONS ON THE EXACT RESPONSE MODIFICATION PROCEDURES

The analysis presented in the current section is the central issue in judging the applicability of the exact formula for receptance modification analysis. The arguments follow the earlier analysis by the author (Brandon (1984c).

As has been commented earlier, the theory of modification analysis for response predictions is well established (see Hammill and Andrew (1973) and Mahalingam (1975a,b)). The condition for the applicability of the exact response formula, equations (5.5) and (5.6) is that W is non-singular. Apparently, therefore, the allowable modifications are otherwise unrestricted in scale. For example, Neves (1978) and Sadeghipour (1984) have used this method for prediction of the effects of large modifications. In the analysis which follows it will be shown that there is a practical bound, not dissimilar to equation (5.12), which should be observed for prudent modification analysis.

The restriction on δv to guarantee non-singularity of W is well known in the linear algebra literature (see for example Noble and Daniel (1979) p168). The condition is

$$\| \; \delta v \; V \; R \; U \; \|_2 \; < \; 1 \qquad (5.13)$$

or

$$\delta v \; < \; \frac{1}{\| \; V \; R \; U \; \|_2} \qquad (5.14)$$

As has been stated, equation (5.14) gives a sufficient condition for the non-singularity of W but by no means a necessary one. As δv increases W has at most $r - 1$ additional singular values beyond the limit of equation (5.14). It is reasonable to enquire what adverse consequences may occur if the limit of equation (5.14) is infringed and δv is increased to an arbitrarily large value. The analysis uses the insight into Weissenburger's method described in chapter 3, in particular the effects of the bracketing theorems described in section 2.11. As with Weissenburger's method the analysis presented here will consider initially the effect of a rank one modification. This will then be generalised to apply to the rank r modification considered here.

5.4.1 Receptance Changes for Unit Rank Modifications

The unit rank modification can be expressed in the form

$$\Delta S = \delta v \; u \; v^t \qquad (5.15)$$

where vectors u and v represent the distribution of the modification and δv allows variation in the magnitude. In this case equation (5.6) results in a 1×1 matrix w

$$w \; = \; I_1 + \; \delta v \; v^t \; R \; u \qquad (5.17)$$

The inverse of this matrix is

$$w^{-1} = \left\{ \frac{1}{1 + \delta v \; v^t \; R \; u} \right\} \qquad (5.18)$$

The condition for non-singularity of w can be established immediately

$$\delta v \; \neq \; - \; \frac{1}{v^t \; R \; u} \qquad (5.19)$$

More important than the non-singularity condition itself is the relationship between the magnitude of w and δv. As with the individual terms of the auxiliary function, as shown in figure 6, the relationship between $\| \; w^{-1} \; \|_2$ and δv is a rectangular hyperbola as shown in figure 14

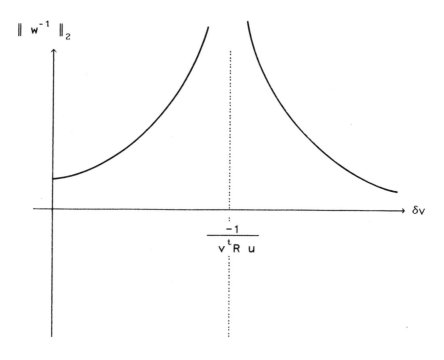

FIGURE 14: HYPERBOLIC FORM OF RECEPTANCE MODIFICATION

Equation (5.19) describes the condition for the existence of a modified receptance matrix. The limiting behaviour establishes the consequences of w becoming singular

$$\| \; R^* \; \|_2 \longrightarrow \infty, \text{ as } \delta v \longrightarrow - \frac{1}{v^t \; R \; u} \qquad (5.20)$$

The physical interpretation of equation (5.20) is that, as δv approaches the limiting value the receptance modification, and hence the modified receptance matrix becomes infinite, exactly the behaviour at resonance. Thus as the magnitude of the proposed modification is increased the adjacent resonance is pulled to the frequency of interest, and, if δv is then allowed to increase further, the resonance passes through the frequency of interest. As was discovered in chapter 3, at most one resonance can pass through the frequency of interest for a given rank one modification.

5.4.2 Receptance Changes for General Modifications

Although the pseudo-force method extends to non-linear systems, (see Mahalingam (1975b)), all of the systems studied in the current volume are linear. The essential property of linearity is the principle of superposition. In this context linearity requires that the effect of a rank r modification is directly equivalent to the application of r individual rank 1 modifications applied sequentially.

Using similar arguments to those described in chapter 3, it can be seen that, for a given distribution, a modification of rank r will result in at most r resonances passing through the frequency of interest, depending on the magnitude of the proposed modification.

It was found in chapter 4 that the degree to which the modes shapes of a modified system differed from the initial system depends, to a considerable degree, on the separation of the resonant frequencies. The closer together the eigenvalues of the initial system, the more complicated are the combinations of the initial eigenvectors in the resulting eigenvectors of the modified system. For present purposes, it is necessary to examine the qualitative behaviour of the receptance matrix if the resonant frequencies of the modes are well separated, when it may be expected that the mode shapes of the modified system are sufficiently similar to those of the initial system for the residual approximations, discussed in section 2.13, to apply.

Consider the behaviour of the receptance matrix initially in the region of the m-th mode. The receptance matrix may be written (following equation (2.39))

$$R = \frac{\phi_m \phi_m^t}{\left(\lambda_m^2 - s^2\right)\phi_m^t M \phi_m} + \text{residuals} \qquad (5.21)$$

After a modification, which causes j resonances to pass through the frequency of interest, the receptance matrix is given by

$$R^* = \frac{\phi_{m+j}^* \phi_{m+j}^{*t}}{\left(\lambda_{m+j}^{2*} - s^2\right)\phi_{m+j}^{t*} M^* \phi_{m+j}^*} + \text{residuals} \qquad (5.22)$$

Thus, whereas the receptance matrix of the initial system is dominated by the m-th mode of the initial system, the (m + j)-th mode of the modified system dominates the receptance matrix.

Viewing this information qualitatively, it can be seen that the modification has suppressed the dominant mode of the initial system, ϕ_m, and amplified the (m+j)-th mode such that it dominates the receptance matrix of the modified system. Consequently the analyst is seeking to predict the behaviour of the modified system, in the region of a resonance, from the behaviour of a different (though closely related) system in the region of a different resonance.

From consideration of the above information, it is suggested that a prudent analyst is likely to observe the condition given in equation (5.14) in predicting the effects of modifications on the receptance matrix.

CHAPTER 6
Methods for Response Sensitivities

6.1 RESPONSE SENSITIVITIES IN CONTEXT

This chapter follows the analysis by Brandon (1987a) which expressed the view that, for a restricted class of problems, modal analysis was an unnecessarily complicated approach to the solution of problems of limited scope.

A large number of problems in structural dynamics involve the treatment of specific unacceptable response characteristics of a structure, which are often localised either in frequency or location. One such problem involves the response of a helicopter structure to the excitation due to the main rotor, which commonly has a fixed driving frequency and identifiable location.

Another problem requiring consideration of local excitation and response properties is the chatter sensitivity of machine tools. In this case, interest often focuses on a limited frequency range (usually in the region of the first mode (see Sweeney and Tobias (1969)) and the excitation and response points coincide, in the region of the cutting zone.

Whereas techniques of sensitivity analysis for modal properties are well developed, response sensitivities have generally been computed indirectly, by evaluating the modal sensitivities first, then using the spectral form of the receptance matrix, and exploiting linearity, given by equation (2.26) or (2.27), to express the response sensitivities as the summation of a series of modal contributions. Researchers who have taken this approach include Vanhonacker (1980,1981), Tomita and Frohrib, (1983) and Yoshimura, (1983, 1987).

In this chapter the response sensitivities will be established in two ways, neither of which requires the explicit evaluation of modal properties. Firstly, the binomial form, described in section 5.3, will be used to derive the coefficients of the Taylor series by term-by-term comparison, in a similar way to the variational derivation of the modal sensitivities in chapter 3. The second strategy will examine the behaviour of Kron's equation in its infinitesimal limiting form.

6.2 DERIVATION OF RESPONSE SENSITIVITIES BY BINOMIAL EXPANSION

Recall initially equation (5.10)

$$R^* = \left(I - (R \; \Delta S) + (R \; \Delta S)^2 - \; ... \; \right) R$$

subject to the convergence criterion, expressed in the least squares norm

$$\| \; R \; \Delta S \; \|_2 \; < \; 1$$

Writing $\delta R = (R^* - R)$, $\Delta S = \delta v \; U \; V$, as in chapters 4 and 5, leads to an expression for δR

$$\delta R \; = \; - \; \delta v \left(\; (R \; U \; V) - \delta v \; (R \; U \; V \;)^2 + \; ... \; \right) R \qquad (6.1)$$

The derivative is obtained by considering the limit

$$\frac{dR}{dv} = \lim_{\delta v \to 0} \frac{\delta R}{\delta v} = R\ U\ V\ R \qquad (6.2)$$

Consistent with the previous methods described in this volume, the sensitivity of the receptance matrix depends only on the existing receptance matrix and the proposed modification. Computationally, equation (6.2) can be evaluated from the middle outwards. Since U is n×r, R U is also n×r and, similarly, V R is r×n. For symmetric modifications (eg for viscously damped non-gyroscopic systems) $R\ U = (\ V\ R\ )^t$. Thus low rank modifications result in computation with matrices of low order.

The convergence criterion is given by

$$\|\ \delta v\ R\ U\ V\ \|_2\ <\ 1 \qquad (6.3)$$

which is similar to the conditions described in chapter 5. The Taylor series may be expected to provide a satisfactory first order approximation if

$$\delta v\ \ll\ \frac{1}{\|\ R\ U\ V\ \|_2} \qquad (6.4)$$

6.3 RESPONSE SENSITIVITIES USING CLOSED FORM MODIFICATION FORMULAE

The derivative of the receptance matrix may be derived alternatively using the exact formula, equations (5.5) and (5.6), described in section 5.2.

Rearranging equation (5.5), and setting $\delta R = R^* - R$, gives:

$$\delta R\ =\ -\ \delta v\ R\ U\ W^{-1}\ V\ R \qquad (6.5)$$

Noting that $\quad \underset{\delta v \longrightarrow 0}{\lim} \left(W^{-1} \right) = I_r$

leads to the same expression for the derivative of R with respect to the proposed design modification:

$$\frac{dR}{dv} = \underset{\delta v \longrightarrow 0}{\lim} \frac{\delta R}{\delta v} = R\ U\ V\ R \qquad (6.2)$$

6.4 SENSITIVITIES WITH RESPECT TO SIMPLE DESIGN MODIFICATIONS

Consider, for example, the effects of a mass addition at the r-th coordinate. In this case δv is given by $s^2\ \delta m$ and the distribution of the change may be written as $U = e_r$, $V = e_r^t$. The standard chain rule of differentiation applies, ie

$$\frac{dR}{dm} = \frac{dR}{dv}\frac{dv}{dm} = s^2\ R\ U\ V\ R = s^2\ c\ c^t \qquad (6.6)$$

where c_r is defined, in the same way as in section 5.2, as the r-th column of R. Thus the receptance sensitivity to a lumped mass change depends only on the column of the receptance matrix corresponding to the position of the modification coordinate. The receptance sensitivities to earthed stiffness links or dampers are given by:

$$\frac{dR}{dc} = s\ R\ U\ V\ R = s\ c\ c^t \qquad (6.7)$$

$$\frac{dR}{dk} = R\ U\ V\ R = c\ c^t \qquad (6.8)$$

With stiffness links or dampers, which link two coordinates, the change is still of rank one, ie the sensitivity may be described in terms of the outer product of two vectors, but now the vectors are composed of a summation of contributions from the corresponding columns of the receptance matrix. For example, the stiffness sensitivity for a simple link is given by

$$\frac{dR}{dk} = R U V R = R\, e_{rs}\, e_{rs}^{t}\, R \qquad (6.9)$$

following the analysis and notation of section 3.2.2, ie e_{rs} has its r-th term +1 and its s-th term -1.

Similarly the response sensitivity with respect to damping is given by

$$\frac{dR}{dc} = s R U V R = s\, R\, e_{rs}\, e_{rs}^{t}\, R \qquad (6.10)$$

CHAPTER 7
Synoptic Review

7.1 APPLICABILITY AND LIMITATIONS OF THE METHODS

As was described in chapter 1, the methods described in this volume are intended for a specific, restricted range of applications. This is the appraisal of the effects of a proposed modification in the absence of design information, using data derived from experimental tests. Each of the methods predicts the structural properties of the modified system using only the corresponding properties of the initial structure and the proposed modification.

In the present section the methods will be reviewed and compared, identifying particularly their relative strengths, weaknesses and limitations.

7.1.1 Exact Modal Reanalysis: Applicability and Limitations

The methods of Weissenburger (1966, 1968) and Pomazal (1969) may be considered as special cases of the more general substructure method (reviewed by Craig (1985)) in a similar way to the exact response methods of chapter 5. As with the exact response methods, considerable benefits (potentially) accrue by taking the approach described here.

Strengths

° Applicable to systems with general damping distribution, including those with a defective eigensystem

° Capable of handling design modifications of arbitrary scale

Weaknesses

° Restricted to modifications of single rank, despite the fact that, for example (as described by Ewins (1984)), a mass modification would affect terms of the mass matrix corresponding to the coordinates of three perpendicular directions

° Although simple in comparison to substructure methods the exact method for modal modification analysis is more complex and computationally intensive than the other three methods

° If objective of the analysis is a response prediction this must be constructed, using the spectral decomposition, using the modified modal characteristics

7.1.2 Modal Sensitivity Analysis: Applicability and Limitations

Modal sensitivity analysis is currently the dominant method in structural modification analysis. It is currently implemented widely in modal analysis software. There is some evidence, however, that a number of workers expect too much of the methods, particularly in the appraisal of gross structural modifications.

Strengths

° Suitable for assessment of the effects of distributed modifications

○ Straightforward to compute using measured
properties of the system

○ Sensitivity coefficients evaluated are not
frequency dependent

Weaknesses

○ Although some authors have described use of
sensitivities for gross modifications, these will
be of doubtful validity for structures with
close eigenvalues and ineffective for modes with
low modal mass

○ Response predictions require complicated
computational procedures if modal sensitivities are
used

7.1.3 Exact Response Reanalysis: Applicability and Limitations

Exact response methods appear to be significantly less
well developed than modal methods. In the context of
"troubleshooting", which is one of the major areas of
applicability of the methods described in the current
volume, the analyst may, however, be interested in an
extremely restricted undesirable structural property. The
response methods are ideal in this case.

Strengths

○ Significantly less computation is required than
for the exact modal method if response at single
frequency, or very limited range of frequencies, is
required

○ Most direct method if the objective of the
analysis is to predict responses

   ° Analytically unrestricted in scale (subject to a simple non-singularity condition)

Weaknesses

   ° Requires evaluation of new response properties at each frequency of interest- quickly becomes uneconomic if responses across a large range of frequency are required

   ° Method requires the inversion of a matrix of the same rank as the modification matrix- practically restricted to modifications of low rank

   ° For guaranteed non-singularity, condition requires a bound on the scale of modification similar to the condition for convergence of the binomial expansion

7.1.4 Response Sensitivities: Applicability and Limitations

Whilst indirect evaluation of response sensitivities is widely practised, through computation of modal sensitivities and exploitation of linearity to use the spectal decomposition, the direct method for receptance sensitivities is not widely known.

Strengths

   ° Significantly less computation is required than for the modal sensitivity method, if response at single frequency, or very limited range of frequencies, is required

   ° Not restricted in distribution

   ° Most direct method if the objective of the analysis is to predict responses

Weaknesses

  ∘ Requires evaluation of new response properties at each frequency of interest- quickly becomes uneconomic if responses across a large range of frequency are required

7.2 STRATEGIC PERSPECTIVES

This text has been, deliberately, restricted in scope, addressing a sub-domain of structural reanalysis, ie that which needs no detailed a-priori model. Because the methods rely on such a simplistic model they should be treated with a significant degree of caution. Nevertheless, it is often necessary, in the analysis of engineering systems, to make judgements and take action based on incomplete data because of the possibility that to wait for more detailed analysis may be even more dangerous! It is hoped that this text will form a sound basis on which such judgements can be made.

This subject, as with all topics in dynamics, has seen an exponential increase in published methods. Much of this literature is original, in that the authors were honestly unaware of previous methods, but not new, as the method described may have already been published previously, albeit in different notation. It is hoped that the approach taken here will lead to further critical evaluation of the literature and rationalisation of methods used.

A further aspiration of the current work is that the balance of research will tip towards direct methods for response analysis rather than indirectly, through initial modal analysis.

References

H M Adelman and R T Hatfka, (Editors), (1987) Sensitivity Analysis in Engineering, Proceedings of a Symposium, NASA Langley Research Center, September 1986.

J S Arora, (1976), Survey of structural reanalysis techniques, Proc ASCE, Journal of the Structural Division, 102 (ST4), pp783-802.

J F Baldwin and S G Hutton, (1985), Natural modes of modified structures, AIAA Journal, 23, pp1737-1743.

M Baruch, (1984), Methods of reference basis for identification of linear dynamic structures, AIAA Journal, 22, pp561-4.

M Baruch and Bar Ishtak, (1978), Optimal weighted orthogonalisation of measured modes, AIAA Journal, 18, pp346-51.

A Berman, (1975), Determining structural parameters from dynamic testing, Shock and Vibration Bulletin, 7, pp10-17.

A Berman, (1980), Improved orthogonality check for measured modes, AIAA Journal, 18, pp1151-2.

A Berman, (1984),Limitations on the identification of discrete structural dynamic models, 2nd International Conference Recent Advances in Structural Dynamics, Institute of Sound and Vibration Research, Southampton, pp427-435.

A Berman and W G Flannelly, (1971), Theory of incomplete models of dynamic structures, AIAA Journal, 9, pp1481-7.

A Berman and E J Nagy, (1983), Improvement of a large analytical model using test data, AIAA Journal, 21, pp1168-73.

A Berman, F S Wei and K V Rao, (1980), Improvement of analytical dynamic models using modal test data, 21st Structures, Structural Dynamics and Materials Conference, Seattle, May pp809-14.

R E D Bishop, G M L Gladwell and S Michaelson, (1965), The Matrix Analysis of Vibration, Cambridge University Press.

R E D Bishop and D C Johnson, (1960), The Mechanics of Vibration, Cambridge University Press.

R E D Bishop and D C Johnson, (1963), An investigation into the theory of resonance testing, Proc Royal Society, A255, pp241-280.

J A Brandon, (1984a), Discussion of alternative Duncan formulations of the eigenproblem for the solution of nonclassically, viscously damped linear systems, Tr ASME Journal of Applied Mechanics, 51, pp904-6.

J A Brandon, (1984b), Derivation and significance of second order modal design sensitivities, AIAA Journal, 22, pp723-4.

J A Brandon, (1984c), Limitations of an exact formula for receptance reanalysis, Int Journal of Numerical Methods in Engineering, 20, pp1575-80.

J A Brandon, (1985), Reply to A H Flax, AIAA Journal, 23, p479.

J A Brandon, (1987a), Eliminating indirect analysis- the potential for receptance sensitivities, Int Journal of Analytical and Experimental Modal Analysis, 2, pp73-75.

J A Brandon, (1987b), The significance and practice of rank estimation in structural dynamics identification, Numerical Methods in Engineering: Theory and Applications, G Pande and J Middleton, (Eds) Martinus Nijhoff, S5,

pp1-8.

J A Brandon, (1988), On the robustness of algorithms for the computation of the pseudo-inverse for modal analysis, 6th International Modal Analysis Conference, Orlando, Florida, pp397-400.

J A Brandon and A Cowley, (1983), A weighted least squares method for fitting circles to frequency response data, Journal of Sound and Vibration, 89, pp419-424.

J A Brandon, K Sadeghipour and A Cowley, (1988), Exact reanalysis techniques for predicting the effects of modification on the dynamic behaviour of structures, their potential and limitations, Int J Machine Tool Design and Manufacture, Vol 28, 4, pp351-7.

J A Brandon, (1989), Derivation and application of the Choleski decomposition for the positive semi-definite matrices used in structural dynamics reanalysis, in Modern Practice in Stress and Vibration Analysis, Editor J E Mottershead, Pergamon, pp225-233.

T K Caughey, (1960), Classical normal modes in damped linear dynamic systems, Tr ASME, Journal of Applied Mechanics, June pp269-271.

T K Caughey and M E J O'Kelly, (1965), Classical normal modes in damped linear systems, Tr ASME, Journal of Applied Mechanics, 32, pp583-588.

J C Chen and J A Garba, (1979), Matrix perturbations for analytical model improvement, 20th Structures, Structural Dynamics and Materials Conference, St Louis, pp428-36.

A R Collar and A Simpson, (1987), Matrices and Engineering Dynamics, Ellis Horwood, Chichester.

J D Collins, J P Young and L Kiefling, (1972), Methods and applications of system identification in shock noise and vibration, Proc Symposium on Identification of Vibrating Structures, ASME Winter Annual Meeting, pp45-71.

R R Craig, (1985), A review of time-domain and frequency-domain component mode synthesis method, in: D R Martinez and A K Miller, Editors, Combined Experimental/ Analytical Modeling of Dynamic Structural Systems, ASME Volume AMD-67, pp1-30.

B Dobson, D J Ewins, D A C Parkes and J Sidhu, (1984,June) Comparison of predicted and measured modal properties, Journal of the Society of Environmental Engineers, 23-2, pp3-11.

K B Elliott and L D Mitchell, (1987), Structural modification using beam elements, 5th International Modal Analysis Conference, London, pp956-965.

D J Ewins, (1980), On predicting point mobility plots from measurements of other mobility parameters, Journal of Sound and Vibration, 70, pp69-75.

D J Ewins, (1984), Modal Analysis: Theory and Applications, Research Studies Press.

W G Flannelly and A Berman, (1972), The state of the art of system identification of aerospace structures, Proc Symposium on System Identification of Vibrating Structures, ASME Winter Annual Meeting, pp121-131.

A H Flax, (1985), Comment on "Derivation and significance of second order modal design sensitivities", AIAA Journal, 23, p478.

K A Foss, (1956), Coordinates which uncouple the equations of motion of damped linear systems, Technical Report 25-30, Massachusetts Institute of Technology.

G L Fox, (1981), A method for estimating the error induced by the Guyan reduction, Shock and Vibration Bulletin, 51, pp19-24.

L Fox, (1952), Escalator methods for latent roots, Quarterly Journal of Mechanics and Applied Mathematics, 5, pp178-90.

R L Fox and M P Kapoor, (1968), Rates of change of eigenvalues and eigenvectors, AIAA Journal, 6 (12) pp.246-9.

R A Frazer, W J Duncan and A R Collar, (1938), Elementary Matrices, Cambridge University Press.

B S Gabri and J T Matthews, (1980), Normal mode testing using multiple exciters under digital control, Journal of the Society of Environmental Engineers, pp25-29.

B E Gage, (1986), Data handling, Workshop on Ground Vibration Testing of Aerospace Structures, Royal Aeronautical Society.

S Garg, (1973), Derivatives of eigensolutions for a general matrix, AIAA Journal, 11 (8), pp.1191-4.

R J Guyan, (1965),Reduction of stiffness and mass matrices, AIAA Journal, 3, p380.

A L Hale and L V Warren, (1983), Concepts of a general substructuring system for structural dynamic analysis, ASME paper 83-DET-19, Design and Production Engineering Technical Conference, Dearborn, Michigan, September 11-14.

J O Hallquist, (1976), An efficient method for determining the effects of mass modifications in damped systems, Journal of Sound and Vibration, 44, pp449-459.

W J Hammill and C Andrew, (1973), Vibration reduction with continuous damping inserts, 13th International Machine Tool Design and Research Conference, Manchester College of Technology, pp495-507.

R D Henshell and J H Ong, (1975), Automatic masters for eigenvalue economisation, Earthquake Engineering and Structural Dynamics, 3, pp375-383.

B Irons, (1963), Eigenvalue economisers in vibration problems, Journal of the Royal Aeronautical Society, 67, pp526-8.

B Irons, (1965), Structural eigenvalue problems, elimination of unwanted variables, AIAA Journal, 3,

pp961-2.

B Irons, (1981), Dynamic reduction of structural models (discussion on a paper by C A Miller), Proc ASCE, Journal of the Structural Division, pp1023-4.

A Jennings, (1977), Matrix Computation for Engineers and Scientists, John Wiley.

A L Klosterman, (1971), On the experimental determination and use of modal representations of dynamic characteristics, PhD Thesis, U Cincinnati.

G Kron, (1963), Diakoptics: The Piecewise Solution of Large Scale Systems, Macdonald.

P Lancaster, (1966), Lambda Matrices and Vibrating Systems, Pergamon Press, Oxford.

P Lancaster, (1960), Free vibrations of lightly damped systems by perturbation methods, Quarterly Journal of Mechanics and Applied Mathematics, 13, pp138-155.

C Lanczos, (1957), Applied Analysis, Pitman.

A Y-T Leung, (1978), An accurate method of dynamic condensation in structural analysis, International Journal Numerical Methods in Engineering, 12, pp1705-15.

A Y-T Leung, (1979), An accurate method of dynamic substructuring with simplified computation, International Journal Numerical Methods in Engineering, 14, pp1241-56.

R C Lewis and D L Wrisley, (1950), A system for the excitation of pure natural modes of complex structures, Journal of the Aeronautical Sciences, 17(11), pp705-722.

A P Lincoln, (1977), Modelling of structural behaviour from frequency response data, Technical Report No 83, Institute of Sound and Vibration Research, Southampton University.

S Mahalingam, (1975a), The synthesis of vibrating systems by use of internal harmonic receptances, Journal of Sound and Vibration, 40, pp337-50.

S Mahalingam, (1975b), The response of vibrating systems

with coulomb and linear damping inserts, Journal of Sound
and Vibration, 41, pp311-320.

D R Martinez and A K Miller, Editors, (1985), Combined
Experimental/ Analytical Modeling of Dynamic Structural
Systems, ASME Volume AMD-67

L Meirovitch, (1967), Analytical Methods in Vibrations,
Macmillan.

L Meirovitch, (1980), Computational Methods in Structural
Dynamics, Sijthoff & Noordhoff.

L Meirovitch, (1986), Elements of Vibration Analysis,
McGraw-Hill.

C A Miller, (1980), Dynamic reduction of structural
models, Proc ASCE, Journal of the Structural Division,
pp2097-2108.

J Morris, (1947), The escalator method, Chapman Hall.

D V Murthy and R T Hatfka, (1987), Survey of methods for
calculating sensitivity of general eigenproblems,
in H M Adelman and R T Hatfka (1987), pp177-196

H G Natke and H Schulze, (1981), Parameter adjustment of
an offshore platform from estimated eigenfrequencies data,
Journal of Sound and Vibration, 77, pp271-285.

R B Nelson, (1976), Simplified calculation of eigenvector
derivatives, AIAA Journal,14, pp1201-5.

F J R Neves, (1978), A study of the damped vibration
behaviour of spindle bearing systems, PhD Thesis,
University of Manchester Institute of Science and
Technology.

N Niedbal, (1985), Experimental system identification for
experimental/analytical correlation and modelling, in: D R
Martinez and A K Miller, (1985) pp195-204.

B Noble, (1969), Applied Linear Algebra, Prentice Hall.

B Noble and J W Daniel, (1977), Applied Linear Algebra,
Prentice Hall.

M Paz, (1984), Dynamic condensation, AIAA Journal, 22,

pp724-7.

R H Plaut and K Huseyin (1976), Derivatives of eigenvalues and eigenvectors in non-self-adjoint systems, AIAA Journal,14 (9) pp.120-5.

R J Pomazal, (1969), The effect of local modifications on the eigenvalues and eigenvectors of damped linear systems, PhD Thesis, Michigan Technological University.

R J Pomazal and V W Snyder, (1970), Local modifications of damped linear systems, AIAA Journal, 9, pp2216-2221.

J I Pritchard, H M Adelman, and R T Hatfke, (1987), Sensitivity derivatives and optimisation of nodal point locations for vibration reduction, in H M Adelman and R T Hatfka, (1987), pp215-231.

Lord Rayleigh, (1945), Theory of Sound, Vol 1, Dover.

L C Rogers, (1970), Derivatives of eigenvalues and eigenvectors, AIAA Journal,8 (5) pp.943-4.

R G Ross, (1971), Synthesis of stiffness and mass matrices from experimental vibration modes, Society of Automotive Engineers,paper 710787.

K Sadeghipour, (1984), Mathematical modelling and sensitivity analysis of complex vibrating systems, PhD Thesis, University of Manchester Institute of Science and Technology.

K Sadeghipour, J A Brandon and A Cowley, (1985), The receptance modification strategy of a complex vibrating system, International Journal of the Mechanical Sciences, 27, pp841-6.

N S Sehmi, (1985), A Newtonian procedure for the solution of the Kron characteristic value problem, Journal of Sound and Vibration, 100, pp409-421.

V N Shah and M Raymund, (1982), Analytical selection of masters for the reduced eigenvalue problem, International Journal Numerical Methods in Engineering, 18, pp89-98.

A Simpson, (1980), The Kron methodology and practical

algorithms for eigenvalue, sensitivity and response analyses of large scale structural systems, Aeronautical Journal, 84 , pp417-433.

A Simpson and B Tabarrok, (1968), On Kron's eigenvalue procedure and related methods of frequency analysis, Quarterly Journal of Mechanics and Applied Mathematics, XXI (1), pp1-39.

G W Skingle and D J Ewins, (1989), Sensitivity analysis using resonance and anti-resonance frequencies- a guide to structural modification, European Forum on Aeroelasticity and Structural Dynamics, Aachen.

G W Stewart, (1973), Introduction to matrix computation, Academic Press.

G Sweeney and S A Tobias, (1969), Survey of basic machine tool chatter research, International Journal of Machine Tool Design and Research 9, pp217-238.

J Tlusty and T Morikawa, (1976), Experimental and computational identification of dynamic structural models, Annals of the CIRP, 25/2/1976, pp497-503.

J Tlusty, F Ismail and E K Prossler, (1978), Identification, modelling and modification of structures defined from measured data, Technical Report, Laboratorium fur Werkzeugmaschinnen und Betriebslehre, Technische Hochschule Aachen.

W M To and D J Ewins, (1989), Structural modification analysis using Rayleigh quotient iteration, in Modern Practice in Stress and Vibration Analysis, Editor J E Mottershead, Pergamon, pp1-9.

K Tomita and D A Frohrib, (1983), Sensitivity functions as predictors of the influence of design parameters on natural response properties, ASME Paper 83-DET-97.

R W Traill-Nash, (1958), On the excitation of pure natural modes in aircraft resonance testing, Journal of the Aeronautical Sciences, 25(12), pp775-778.

G L Turner, M G Milstead and P Hanks, (1985), The adaptation of Kron's method for use with large finite element models, ASME paper 85-DET-183, Design Engineering Division Conference on Mechanical Vibration and Noise, Cincinnati, Ohio.

S Utku, J L M Clemente and M Salama, (1985), Errors in reduction methods, Computers and Structures, 21, pp1153-7.

P Vanhonacker, (1980), The use of modal parameters of mechanical structures in sensitivity analysis, system synthesis and system identification methods, PhD Thesis, University of Leuven.

P Vanhonacker, (1981), An introduction to sensitivity analysis, Sixth Modal Analysis Seminar, University of Leuven.

P van Loon, (1974), Modal parameters of mechanical structures, PhD Thesis, K U Leuven.

J Wang, W Heylen and P Sas, (1987), Accuracy of structural modification techniques, 5th International Modal Analysis Conference, London, pp65-71.

J T Weissenburger, (1966), The effect of local modifications on the eigenvalues and eigenvectors of linear systems, ScD Dissertation, Sever Institute, Washington U, St Louis, Mo.

J T Weissenburger, (1968), The effect of local modifications on the vibration characteristics of linear systems, Tr ASME, Journal of Applied Mechanics, 35, pp327-332.

J E Whitesell, (1980), Design sensitivity in mechanical systems, PhD Thesis, Michigan State U.

A R Whittaker and M M Sadek, (1980), Optimisation of machine tool structures using structural modelling techniques, 1st International Conference Recent Advances in Structural Dynamics, Southampton, pp211-223.

J H Wilkinson, (1965), The Algebraic Eigenvalue

Problem, Oxford University Press.

J H Wilkinson and C Reinsch, (Eds.),(1971), Handbook for Automatic Computation, Vol.2 Linear Algebra, Springer Verlag.

W H Wittrick, (1962), Rates of change of eigenvalues with reference to buckling and vibration problems, Journal of the Royal Aeronautical Society, 66, pp590-1.

M Yoshimura, (1983), Design sensitivity analysis of frequency response in machine structures, ASME Paper 83-DET-50.

M Yoshimura, (1987), Application of design sensitivity analysis for greater improvement on machine structural dynamics, in Adelman and Hatfka (1987), pp285-298.

APPENDIX 1
The Moore-Penrose Generalised Inverse

A1.1 THE MOORE-PENROSE GENERALISED INVERSE: CONTEXT

Inconsistent and/or underdetermined sets of equations are a common occurrence in structural dynamics, a typical example being the identification problem where the order of the data matrix (which may not be square) rarely matches the number of modes represented in the data. Under such conditions the solution of the set of equations cannot be achieved using the classical inverse of a square matrix.

A unique solution is, however, available in terms of the method of least squares. As will be seen, the Moore-Penrose generalised inverse (also called the pseudo-inverse, although there are other pseudo-inverses used in linear algebra) provides a convenient formulation of the problem, although there are a number of algorithmic approaches, not necessarily all of equal merit, described as the pseudo-inverse. These have been discussed by the author previously (Brandon (1988)).

The pseudo-inverse may be defined according to its applications or its fundamental properties.

In the first case, consider the set of linear algebraic equations:

$$A x = b \qquad\qquad (A1.1)$$

where A is m×n, x is n×1 and b is m×1. Then the unique solution, which minimises $\| A x - b \|_2$ and $\| x \|_2$,is given by

$$x \; = \; A^g \; b \qquad\qquad (A1.2)$$

where A^g is the Moore-Penrose generalised inverse or pseudo-inverse.

The pseudo-inverse A^g of a matrix A is defined alternatively by its symmetry properties as follows

$$A \quad A^g \quad A \quad = \quad A \qquad\qquad (A1.3i)$$

$$A^g \quad A \quad A^g \; = \; A^g \qquad\qquad (A1.3ii)$$

$$A \; A^g \quad \text{and} \quad A^g \quad A \quad \text{are symmetric} \qquad (A1.3iii)$$

A1.2   THE RESTRICTED PSEUDO-INVERSE

The definition of the pseudo-inverse generally made in structural dynamics is restricted in the sense that the matrix to be solved must be of full rank (ie the rank k, ie the number of independent vectors which span the rows/columns of A , is equal to the smaller of m or n). In this case

$$A^g \; = \; A^t \; (\; A \; A^t \;)^{-1} \qquad\qquad (A1.4)$$

$$\text{if } m < n$$

and

$$A^g \; = \; (\; A^t \; A \;)^{-1} \; A^t \qquad\qquad (A1.5)$$

$$\text{if } m > n$$

The verification that these expressions define the pseudo-inverse in the full rank case can be achieved simply by substitution into the above expressions ((A1.3i)-(A1.3iii)) which define the pseudo-inverse.

A1.3 EXTENSION TO THE RANK DEFICIENT CASE

If the matrix A is rank deficient then the matrix to be inverted will be singular. With the choice of reliable software the identification algorithms will indicate this. The fact that the restricted pseudo-inverse cannot be computed, in the rank deficient case, by equations (A1.4) or (A1.5), does not mean that the pseudo-inverse itself does not exist, merely that another method must be employed for its evaluation. The prevailing method in structural dynamics for deriving the pseudo-inverse is the singular value decomposition of the matrix, which is widely acknowledged as of great computational value for unreliable data.

The singular value decomposition is often defined in an analogous way to equations (A1.4) and (A1.5) ie

$$A = U \quad S \quad V^H \tag{A1.6}$$

where the columns of U mxm comprise the normalised (euclidean) eigenvectors of $A A^t$, the columns of V^H nxn are composed of the eigenvectors of $A^t A$ and S has the square roots of the non-zero eigenvalues of $A A^t$ (the singular values) on its leading diagonal (elsewhere all zeroes).

The pseudo-inverse can be defined in terms of the same factorisation

$$A^g = V \quad S^g \quad U^H \tag{A1.7}$$

where U^H is the Hermitian transpose of U (ie complex numbers are both transposed and conjugated) and S^g has the same structure as S but the entries on the leading diagonal are the reciprocals of the singular values of

A.

Recalling the orthogonality properties of eigenvectors ie $U\ U^t = I_m$ and $V\ V^t = I_n$, substitution in (A1.3i)-(A1.3iii) verifies that this is an alternative formulation of the pseudo-inverse. In contrast to the restricted pseudo-inverse this factorisation is tolerant of rank deficient data matrices.

There is an extensive literature on the computation of the singular value decomposition. Reliable procedures are available in packages such as NAG, EISPAC and LINPAC. It is generally regarded that the singular value decomposition gives the most reliable means of estimating the rank of a data matrix.

APPENDIX 2

Derivation of Differential Form of the Modal Sensitivities

Consider the non-self adjoint eigenvalue problem given by

$$(A + \lambda_i B)\phi_i = 0 \qquad (A2.1)$$

$$(A^t + \lambda_i B^t)\theta_i = 0 \qquad (A2.2)$$

and consider a change in the design variable v, which may be a combination of elements of both A and B (corresponding to the realistic physical context where it is not in general possible to make a modification of stiffness which has no implications for the mass distribution).

Differentiating (A2.1) with respect to v

$$\left(\frac{\partial A}{\partial v} + \frac{\partial \lambda_i}{\partial v} B + \lambda_i \frac{\partial B}{\partial v} \right) \phi_i + \left(A + \lambda_i B \right) \frac{\partial \phi_i}{\partial v} = 0$$
$$(A2.3)$$

Premultipling by θ_i^t and exploiting the orthogonality of eigenvectors and (A2.2) gives

$$\frac{\partial \lambda_i}{\partial v} = \frac{- \theta_i^t \left\{ \frac{\partial A}{\partial v} + \lambda_i \frac{\partial B}{\partial v} \right\} \phi_i}{\theta_i^t \, B \, \phi_i}$$

(A2.4)

The eigenvector sensitivities are found by premultipling equation (A2.3) by θ_j^t (for $j \neq i$) and again exploiting the orthogonality of eigenvectors

$$\frac{\partial \phi_i}{\partial v} = \sum_{j=1, j \neq i}^{n} \alpha_{ij} \, \phi_j$$

(A2.5)

where

$$\alpha_{ij} = \frac{- \theta_j^t \left\{ \frac{\partial A}{\partial v} + \lambda_i \frac{\partial B}{\partial v} \right\} \phi_i}{\theta_j^t \left\{ A + \lambda_i \, B \right\} \phi_j}$$

(A2.6)

Using $\theta_j^t A \phi_j = \lambda_j \theta_j^t B \phi_j$ gives

$$\alpha_{ij} = \frac{- \theta_i^t \left\{ \frac{\partial A}{\partial v} + \lambda_i \frac{\partial B}{\partial v} \right\} \phi_i}{\left(\lambda_i - \lambda_j \right) \theta_j^t \, B \, \phi_j}$$

(A2.7)

Differentiating (A2.3), and assuming that the modification is linear in A and B (ie $\frac{\partial^2 A}{\partial v^2} = \frac{\partial^2 B}{\partial v^2} = 0$) gives

$$\left(\frac{\partial^2 \lambda_i}{\partial v^2} B + 2 \frac{\partial \lambda_i}{\partial v} \frac{\partial B}{\partial v} \right) \phi_i$$

$$+ 2 \left(\frac{\partial A}{\partial v} + \frac{\partial \lambda_i}{\partial v} B + \lambda_i \frac{\partial B}{\partial v} \right) \frac{\partial \phi_i}{\partial v}$$

$$+ \left(A + \lambda_i B \right) \frac{\partial^2 \phi_i}{\partial v^2} = 0$$

$$(A2.8)$$

The second order eigenvalue sensitivities are obtained, as before, by premultiplying by θ_i^t and using the orthogonality conditions:

$$\frac{\partial^2 \lambda_i}{\partial v^2} = \frac{- 2 \left[\frac{\partial \lambda_i}{\partial v} \theta_i^t \frac{\partial B}{\partial v} \phi_i + \theta_i^t \left\{ \frac{\partial A}{\partial v} + \lambda_i \frac{\partial B}{\partial v} \right\} \right] \frac{\partial \phi_i}{\partial v}}{\theta_i^t B \phi_i}$$

$$(A2.9)$$

In an analogous way, the second order eigenvector sensitivities are obtained by premultiplying by θ_j^t, once again using the orthogonality conditions:

$$\frac{\partial^2 \phi_i}{\partial v^2} = \sum_{j=1, j \neq i} \beta_{ij} \phi_j$$

$$(A2.10)$$

$$\beta_{ij} = \frac{- 2 \left[\frac{\partial \lambda_i}{\partial v} \theta_j^t \frac{\partial B}{\partial v} \phi_i + \theta_j^t \left\{ \frac{\partial A}{\partial v} + \frac{\partial \lambda_i}{\partial v} B + \lambda_i \frac{\partial B}{\partial v} \right\} \frac{\partial \phi_i}{\partial v} \right]}{\left(\lambda_i - \lambda_j \right) \theta_j^t B \phi_j}$$

$$(A2.11)$$

APPENDIX 3

Proof of Kron's Theorem

Kron's theorem was expressed in section 5.2 in the following terms:

For a dynamic stiffness matrix S, with associated receptance matrix $R = S^{-1}$, if a localised modification ΔS the form $\delta\, U\, V$, where δ is a variable scalar, U is n×r and V is r×n, then the new receptance matrix $R^* = \left(S + \Delta S\right)^{-1}$ is given by

$$R^* = R - \delta\, R\, U\, W^{-1}\, V\, R \qquad\qquad (A3.1)$$

where

$$W = I_r + \delta\, V\, R\, U \qquad\qquad (A3.2)$$

subject to the condition $|\,W\,| \neq 0$ (ie W non-singular). The proof of the theorem only requires multiplying R^* by $\left(S + \Delta S\right)$ to give a result I_n, ie it is required to prove

$$\left(R - \delta\, R\, U\, W^{-1}\, V\, R\right)\left(R^{-1} + \delta\, U\, V\right) = I_n \qquad (A3.3)$$

Expanding the left hand side gives

$$I_n - \delta R U W^{-1} V + \delta R U V - \delta^2 R U W^{-1} V R U V =$$

$$I_n - \delta R U \left(W^{-1} - I_r + \delta W^{-1} V R U \right) V =$$

$$I_n + \delta R U \left(I_r - W^{-1} \left(I + \delta V R U \right) \right) V =$$

$$I_n + \delta R U \left(I_r - I_r \right) \right) V = I_n \qquad \text{QED}$$

Index